The ARRL's
Tech Q&A

• •

Your Quick & Easy Path to
a Technician Ham License

Edited By:
Larry D. Wolfgang, WR1B

Contributors:
Chuck Hutchinson, K8CH
R. Dean Straw, N6BV
R. Jan Carman, K5MA

Editorial Assistants:
Maty Weinberg, KB1EIB
Helen Dalton, KB1HLF

Production Staff:
David Pingree, N1NAS, Senior Technical Illustrator
Jodi Morin, KA1JPA, Assistant Production Supervisor: Layout
Paul Lappen, Production Assistant: Layout
Kathy Ford, Proofreader
Sue Fagan, Graphic Design Supervisor: Cover Design
Michael Daniels, Technical Illustrator
Michelle Bloom, WB1ENT, Production Supervisor: Layout

• •

Published By:

ARRL *The national association for*
AMATEUR RADIO

225 Main Street
Newington, CT 06111-1494

This book may be used for Technician license exams given beginning July 1, 2003. The Technician (Element 2) question pool in this book is expected to be used for exams given until July 1, 2007. (This ending date assumes no FCC Rules changes to the Amateur Radio license structure or privileges for this license class, which would force the VEC Question Pool Committee to modify this question pool.) *QST* and ARRLWeb (www.arrl.org) will have news about any Rules changes.

Contents

Congratulations! You have taken the first step into the exciting world of Amateur Radio! You are about to join nearly 700,000 licensed Amateur Radio operators in the United States and nearly three million people around the world who call themselves "hams." You will soon be communicating with some of those hams — in your community, in your part of the state or around the world.

There are many important reasons why governments around the world allow Amateur Radio operators to use valuable radio frequencies for personal communications. As a licensed Amateur Radio operator, you will become part of a large group of trained communicators and electronics technicians. You will be an important emergency communications resource for your neighbors and fellow citizens. Who knows when you may find yourself in the situation of having the only communications link outside your neighborhood? Whether you are caught in a flood, earthquake or other natural disaster or answering a call from someone else, you can provide the knowledge and resources to help. Don't wait for a real emergency to learn emergency communications methods, though. Practice your skills by participating in the many public service opportunities in your area. You might even want to sign up for ARRL's on-line Emergency Communications Course.

No matter what other interests you may have, you will find ways to tie them to Amateur Radio.

- *Computers?* Hams have been using computers to enhance their enjoyment of their hobby since the 1970s. Computer-based data modes are among the fastest growing methods of Amateur Radio communication.
- *Video?* Using computer software and video cameras, hams exchange moving or still images by using Amateur Television (ATV) or Slow-Scan Television (SSTV).
- *Electronics?* There is no better way to learn about electronics than to prepare for a license exam or to build a useful piece of equipment for your ham station.
- *Hiking, biking or other outdoor activities?* Many hams love nothing more than packing a wire antenna and a small, lightweight radio, and heading for the great outdoors.

Whether across town or across the sea, hams are always looking for new friends. So wherever you may happen to be, you are probably near someone — perhaps a whole club — who would be glad to help you get started. If you need help contacting hams, instructors, Volunteer Examiners or clubs in your area, contact us here at ARRL Headquarters. We'll help you get in touch with someone near you. (See the contact information at the bottom of the next page.)

When you pass that exam and enter the exciting world of Amateur Radio, you'll find plenty of activity to keep you busy. You'll also find plenty of friendly folks who are anxious to help you get started. Amateur Radio has many interesting areas to explore. You may be interested in one particular aspect of the hobby now, but be willing to try something new occasionally. You'll discover a world of unlimited potential!

Most of the active radio amateurs in the United States are members of ARRL. The hams' own organization since 1914, ARRL is truly the national association for Amateur Radio. We provide training materials and other services, and represent our members nationally and internationally. *ARRL's Tech Q & A* is just one of the many ARRL publications for all levels and interests in Amateur Radio. You don't need a ham license to join. If you're interested in ham radio, we're interested in you. It's as simple as that! We have included an invitation for you to join ARRL at the back of this book.

If you have comments or suggestions about this book, please use the Feedback Form at the back of this book. We'd like to hear from you. Thanks, and good luck!

David Sumner, K1ZZ
Executive Vice President
Newington, Connecticut
March 2003

New Ham Desk
ARRL Headquarters
225 Main Street
Newington, CT
06111-1494
(860) 594-0200

Prospective new amateurs call:
800-32-NEW-HAM (800-326-3942)
You can also contact us via e-mail:
newham@arrl.org
or check out **ARRLWeb**:
http://www.arrl.org/

When to Expect New Books

A Question Pool Committee (QPC) consisting of representatives from the various Volunteer Examiner Coordinators (VECs) prepares the license question pools. The QPC establishes a schedule for revising and implementing new Question Pools. The current Question Pool revision schedule is as follows:

Exam	*Question Pool Valid Through*	*Current Study Guides*
Technician (Element 2)	June 30, 2007	*Now You're Talking!*, 5th Edition, *ARRL's Tech Q & A*, 3rd Edition *ARRL's Technician Class Video Course*, 4th Edition
General (Element 3)	June 30, 2004	*The ARRL General Class License Manual*, 4th Edition *ARRL's General Q & A*, 1st Edition *ARRL's General Class Video Course*, 4th Edition
Amateur Extra (Element 4)	June 30, 2006	*The ARRL Extra Class License Manual*, 8th Edition *ARRL's Extra Q & A*, 1st Edition

As new Question Pools are released, ARRL will produce new study materials before the effective dates of the new Pools. Until then, the current Question Pools will remain in use, and current ARRL study materials, including this book, will help you prepare for your exams.

As the new Question Pool schedules are confirmed, the information will be published in *QST* and on **ARRLWeb** at **www.arrl.org**.

How to Use
This Book

To earn a Technician Amateur Radio license, you must pass the Technician written exam, FCC Element 2. You will have to know some basic electronics theory and Amateur Radio operating practices and procedures. In addition, you will need to learn about some of the rules and regulations governing the Amateur Service, as contained in Part 97 of Title 47 of the Code of Federal Regulations — the Federal Communications Commission (FCC) Rules.

The Element 2 exam consists of 35 questions about Amateur Radio rules, theory and practice. A passing score is 74%, so you must answer 26 of the 35 questions correctly to pass. (Another way to look at this is that you can get as many as 9 questions wrong, and still pass the test.)

The questions and multiple-choice answers in this book are printed exactly as they were written by the Volunteer Examiner Coordinators' Question Pool Committee, and exactly as they will appear on your exam. (Be careful, though. The letter positions of the answer choices may be scrambled, so you can't simply memorize an answer letter for each question.) In this book, the letter of the correct answer is printed in **boldface type** just before the explanation. If you want to study without knowing the correct answer right away, simply cover the answer letters with your hand or a slip of paper.

As you read the explanations for many of the questions you will find words printed in **boldface type**. These words are important terms, and will help you identify the correct answer to the question.

ARRL Study Materials

To earn a Technician license with Morse code privileges, you'll have to be able to send and receive the international Morse code at a rate of 5 wpm. This license gives you operating privileges on the high-frequency (HF) bands, and offers the excitement of worldwide communications.

ARRL offers *Your Introduction to Morse Code*, a set of cassette tapes or audio CDs that teach you the code. The two cassettes or audio CDs introduce each of the required charcaters and provide practice on each character as it is introduced. Then the character is used in words and text before proceeding to the next character. After all characters have been introduced, there is plenty of practice at 5 wpm to help you prepare for the Element 1 (5 wpm) exam.

The Morse code on *Your Introduction to Morse Code* is sent using faster characters, with extra space between characters to slow the overall code speed. This same technique is used on ARRL/VEC Morse code exams.

For those who prefer a computer program to learn and practice Morse code, ARRL offers *Ham University* for IBM PC and compatible computers.

Even with the tapes, CDs or computer program, you'll want to tune in the code-practice sessions transmitted by W1AW, the ARRL Headquarters station. A W1AW schedule appears on page 8 of this book. Any updates will be found at **ARRLWeb, www.arrl.org**. For information on how to order any ARRL publication or *Your Introduction to Morse Code*, write to ARRL Headquarters, 225 Main St, Newington, CT 06111-1494, tel (toll free) 1-888-277-5289. You can also order at **ARRLWeb.**

The Technician License

Earning a Technician Amateur Radio license is a good way to begin enjoying ham radio. There is no Morse code exam for this license, and the Element 2 written exam is not difficult. There is no difficult math or electronics background required. You are sure to find the operating privileges available to a Technician licensee to be worth the time spent learning about Amateur Radio. After passing the exam, you will be able to operate on *every frequency* above 50 megahertz that is assigned to the Amateur Radio Service. With full operating privileges on those bands, you'll be ready to experience the excitement of Amateur Radio!

Perhaps you are mainly interested in local communications using FM repeaters. Maybe you want to use your computer to explore the many digital modes of communication. If your eyes turn to the stars on a clear night, you might enjoy tracking the amateur satellites and using them to relay your signals to other amateurs around the world!

Once you make the commitment to study and learn what it takes to pass the exam, you *will* accomplish your goal. Many people pass the exam on their first try, so if you study the material and are prepared, chances are good that you will soon have your license. It may take you more than one attempt to pass the Technician license exam, but that's okay. There is no limit to how many times you can take it. Many Volunteer Examiner Teams have several exam versions available, so you may even be able to try the exam again at the same exam session. Time and available exam versions may limit the number of times you can try the exam at a single exam session. If you don't pass after a couple of tries you will certainly benefit from more study of the question pools before you try again.

An Overview of Amateur Radio

Earning an Amateur Radio license, at whatever level, is a special achievement. The nearly 700,000 people in the US who call themselves Amateur Radio operators, or hams, are part of a global fraternity. Radio amateurs provide a voluntary, noncommercial, communication service. This is especially true during natural disasters or other emergencies. Hams have made many important contributions to the field of electronics and communications, and this tradition continues today. Amateur Radio experimentation is yet another reason many people become part of this self-disciplined group of trained operators, technicians and electronics experts — an asset to any country. Hams pursue their hobby purely for personal enrichment in technical and operating skills, without any type of payment except the personal satisfaction they feel from a job well done!

Radio signals do not know territorial boundaries, so hams have a unique ability to enhance international goodwill. Hams become ambassadors of their country every time they put their stations on the air.

Amateur Radio has been around since the early 1900s. Hams have always been at the forefront of technology. Today, hams relay signals through their own satellites, bounce signals off the moon, relay messages automatically through computerized radio networks and use any number of other "exotic" communications techniques. Amateurs talk from hand-held transceivers through mountaintop repeater stations that can relay their signals to other hams' cars or homes. Hams send their own pictures by television, talk with other hams around the world by voice or, keeping alive a distinctive traditional skill, tap out messages in Morse code. When emergencies arise, radio amateurs are on the spot to relay information to and from disaster-stricken areas that have lost normal lines of communication.

The US government, through the Federal Communications Commission (FCC), grants all US Amateur Radio licenses. This licensing procedure ensures operating skill and electronics know-how. Without this skill, radio operators, because of improperly adjusted equipment or neglected regulations, might unknowingly cause interference to other services using the radio spectrum.

Figure 1—With a Technician Amateur Radio license, you'll be able to use hand-held radios to operate over FM repeaters on the VHF and UHF bands.

Who Can Be a Ham?

The FCC doesn't care how old you are or whether you're a US citizen. If you pass the examination, the Commission will issue you an amateur license. Any person (except the agent of a foreign government) may take the exam and, if successful, receive an amateur license. It's important to understand that if a citizen of a foreign country receives an amateur license in this manner, he or she is a US Amateur Radio operator. (This should not be confused with a reciprocal permit for alien amateur licensee, which allows visitors from certain countries who hold valid amateur licenses in their homelands to operate their own stations in the US without having to take an FCC exam.)

License Structure

Anyone earning a new Amateur Radio license can earn one of three license classes — Technician, General and Amateur Extra. These vary in degree of knowledge required and frequency privileges granted. Higher class licenses have more comprehensive examinations. In return for passing a more difficult exam you earn more frequency privileges (frequency space in the radio spectrum). The vast majority of beginners start with the most basic license, the Technician, although it's possible to start with any class of license.

Technician licensees who learn the international Morse code and pass an exam to demonstrate their knowledge of code at 5 words per minute gain some frequency privileges on four of the amateur high-frequency (HF) bands. This license was previously called the Technician Plus license, and many amateurs will refer to it by that name. **Table 1** lists the amateur license classes you can earn, along with a brief description of their exam requirements and operating privileges. On those four HF bands you will experience the thrill of *working* (contacting) other Amateur Radio operators in just about any country in the world. There's nothing quite like making friends with other amateurs around the world. See **Figure 2**.

Although there are also other amateur license classes, the FCC is no longer issuing new licenses for these classes. The Novice license was long considered the beginner's license. Exams for this license were discontinued as of April 15, 2000. The FCC also stopped issuing new Advanced class licenses on that date. They will continue to renew previously issued licenses, however, so you will probably meet some Novice and Advanced class licensees on the air.

Table 1

Amateur Operator Licenses†

Class	Code Test	Written Examination	Privileges
Technician		Basic theory and regulations. (Element 2)*	All amateur privileges above 50.0 MHz.
Technician with Morse code credit	5 wpm (Element 1)	Basic theory and regulations. (Element 2)*	All "Novice" HF privileges in addition to all Technician privileges.
General	5 wpm (Element 1)	Basic theory and regulations; General theory and regulations. (Elements 2 and 3)	All amateur privileges except those reserved for Advanced and Amateur Extra class.
Amateur Extra	5 wpm (Element 1)	All lower exam elements, plus Extra-class theory (Elements 2, 3 and 4)	All amateur privileges.

†A licensed radio amateur will be required to pass only those elements that are not included in the examination for the amateur license currently held.

*If you have a Technician-class license issued before March 21, 1987, you also have credit for Elements 1 and 3. You must be able to prove your Technician license was issued before March 21, 1987 to claim this credit.

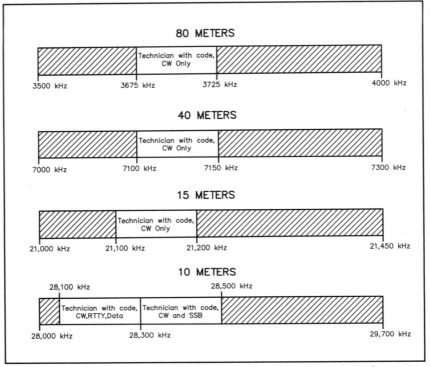

Figure 2.

The written Technician exam, called Element 2, covers some basic radio fundamentals and knowledge of some of the rules and regulations in Part 97 of the FCC Rules. With a little study you'll soon be ready to pass the Technician exam.

Each step up the Amateur Radio license ladder requires the applicant to pass the lower exams. So if you want to start out as a General class or even an Amateur Extra class licensee, you must also pass the Technician written exam. This does not mean you have to pass the Technician exam again if you already hold a Technician license! Your valid Amateur Radio license gives you credit for all the exam elements of that license when you go to upgrade. If you now hold a Technician license, you will only have to pass the Element 1 Morse code exam and the Element 3 General class written exam for a General class license.

A Technician license gives you the freedom to develop operating and technical skills through on-the-air experience. These skills will help you upgrade to a higher class of license, with additional privileges.

As a Technician, you can use a wide range of frequency bands — *all amateur bands above 50 MHz*, in fact. You'll be able to use point-to-point communications on VHF FM, and repeaters, packet radio and orbiting satellites to relay your signals over a wider area. You can provide public service through emergency communications and message handling.

Learning Morse Code

Even if you don't plan to use Morse code now, there may come a time when you decide you would like to upgrade your license and earn some operating privileges on the high-frequency (HF) bands. Learning Morse code is a matter of practice. Instructions on learning the code, how to handle a telegraph key, and so on, can be found in *The ARRL General Class License Manual*, published by the ARRL. In addition, *Your Introduction to Morse Code*, ARRL's package to teach Morse code, is available with two cassette tapes or two audio CDs. *Your Introduction to Morse Code* was designed for beginners, and will help you learn Morse code. You will be ready to pass your 5 word-per-minute code exam when you finish the lessons on *Your Introduction to Morse Code*. You can purchase any of these products from your local Amateur Radio equipment dealer or directly from the ARRL, 225 Main St, Newington, CT 06111. To place an order, call, toll-free, **888-277-5289**. You can also send e-mail to: **pubsales@arrl.org** or check out our World Wide Web site: **www.arrl.org/** Prospective new amateurs can call: **800-32-NEW HAM (800-326-3942)** for additional information.

Besides listening to code tapes or CDs, some on-the-air operating experience will be a great help in building your code speed. When you are in the middle of a contact via Amateur Radio, and have to copy the code the other station is sending to continue the conversation, your copying ability will improve quickly! Although you did not have to pass a Morse code test to earn your Technician license, there are no regulations prohibiting you from using code on the air. Many amateurs operate code on the VHF and UHF bands.

ARRL's Maxim Memorial Station, W1AW, transmits code practice and information bulletins of interest to all amateurs. These code-practice sessions and Morse code bulletins provide an excellent opportunity for code practice. **Table 3** is a W1AW operating schedule.

Station Call Signs

Many years ago, by international agreement, the nations of the world decided to allocate certain call-sign prefixes to each country. This means that if you hear a radio station call sign beginning with W or K, for example, you know the station is licensed by the United States. A call sign beginning with the letter G is licensed by Great Britain, and a call sign beginning with VE is from Canada. *The ARRL DXCC List* is an operating aid no ham who is active on the HF bands should be without. That booklet, available from the ARRL, includes the common call-sign prefixes used by amateurs in virtually every location in the world. It also includes a check-off list to help you keep track of the countries you contact as you work toward collecting QSL cards from 100 or more countries to earn the prestigious DX Century Club award. (DX is ham lingo for distance, generally taken on the HF bands to mean any country outside the one from which you are operating.)

The International Telecommunication Union (ITU) radio regulations

Table 2
Amateur Operating Privileges

US Amateur Bands

April 15, 2000   Novice, Advanced and Technician Plus Allocations

New Novice, Advanced and Technician Plus licenses will not be issued *after* April 15, 2000, but *existing* Novice, Technician Plus and Advanced class licenses are unchanged. Amateurs can continue to renew these licenses. Technicians who pass the 5 wpm Morse code exam *after* that date have Technician Plus privileges, although their license says Technician. They must retain the 5 wpm Certificate of Successful Completion of Examination (CSCE) as proof. The CSCE is valid indefinitely for operating authorization, but is valid only for 365 days for upgrade credit.

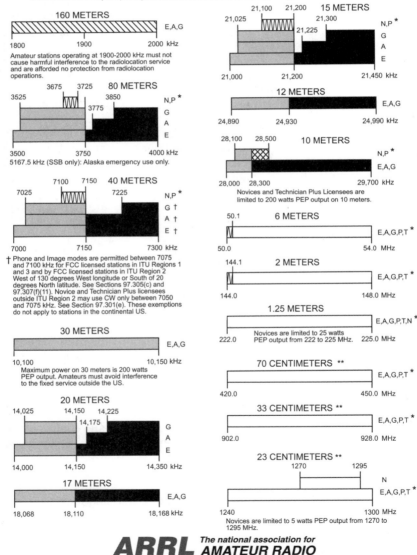

US AMATEUR POWER LIMITS

At all times, transmitter power should be kept down to that necessary to carry out the desired communications. Power is rated in watts PEP output. Unless otherwise stated, the maximum power output is 1500 W. Power for all license classes is limited to 200 W in the 10,100-10,150 kHz band and in all Novice subbands below 28,100 kHz. Novices and Technicians are restricted to 200 W in the 28,100-28,500 kHz subbands. In addition, Novices are restricted to 25 W in the 222-225 MHz band and 5 W in the 1270-1295 MHz subband.

Operators with Technician class licenses and above may operate on all bands above 50 MHz. For more detailed information see *The ARRL's FCC Rule Book.*

—— KEY ——

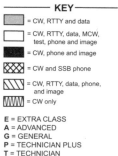

= CW, RTTY and data

= CW, RTTY, data, MCW, test, phone and image

= CW, phone and image

= CW and SSB phone

= CW, RTTY, data, phone, and image

= CW only

E = EXTRA CLASS
A = ADVANCED
G = GENERAL
P = TECHNICIAN PLUS
T = TECHNICIAN
N = NOVICE

*Technicians who have passed the 5 wpm Morse code exam are indicated as "P".

**Geographical and power restrictions apply to all bands with frequencies above 420 MHz. See *The ARRL's FCC Rule Book* for more information about your area.

All licensees except Novices are authorized all modes on the following frequencies:

2300-2310 MHz
2390-2450 MHz
3300-3500 MHz
5650-5925 MHz
10.0-10.5 GHz
24.0-24.25 GHz
47.0-47.2 GHz
75.5-76.0, 77.0-81.0 GHz
119.98-120.02 GHz
142-149 GHz
241-250 GHz
All above 300 GHz

For band plans and sharing arrangements, see *The ARRL's FCC Rule Book.*

outline the basic principles used in forming amateur call signs. According to these regulations, an amateur call sign must be made up of one or two characters (the first one may be a numeral) as a prefix, followed by a numeral, and then a suffix of not more than three letters. The prefixes W, K, N and A are used in the United States. When the letter A is used in a US amateur call sign, it will always be with a two-letter prefix, AA to AL. The continental US is divided into 10 Amateur Radio call districts (sometimes called areas), numbered 0 through 9. **Figure 3** is a map showing the US call districts.

For information on the FCC's call-sign assignment system, and a table listing the blocks of call signs for each license class, see *ARRL's FCC Rule Book.* You may keep the same call sign when you change license class, if you wish. You must indicate that you want to receive a new call sign when you fill out an FCC Form 605 to apply for the exam or change your address.

The FCC also has a vanity call sign system. Under this system the FCC will issue a call sign selected from a list of preferred available call signs. While there is no fee for an Amateur Radio license, there is a fee for the selection of a vanity call sign. The current fee is $14.50 for a 10-year Amateur Radio license, paid upon application for a vanity call sign and at license renewal after that. (That fee may change as costs of administering the program change.) The latest details about the vanity call sign system are available from ARRL Regulatory Information, 225 Main Street, Newington, CT 06111-1494 and on **ARRLWeb** at **www.arrl.org/**

Table 3

W1AW Schedule

PACIFIC	MTN	CENT	EAST	MON	TUE	WED	THU	FRI
6 AM	7 AM	8 AM	9 AM		FAST CODE	SLOW CODE	FAST CODE	SLOW CODE
7 AM-1 PM	8 AM-2 PM	9 AM-3 PM	10 AM-4 PM	VISITING OPERATOR TIME (12 PM-1 PM CLOSED FOR LUNCH)				
1 PM	2 PM	3 PM	4 PM	FAST CODE	SLOW CODE	FAST CODE	SLOW CODE	FAST CODE
2 PM	3 PM	4 PM	5 PM	CODE BULLETIN				
3 PM	4 PM	5 PM	6 PM	TELEPRINTER BULLETIN				
4 PM	5 PM	6 PM	7 PM	SLOW CODE	FAST CODE	SLOW CODE	FAST CODE	SLOW CODE
5 PM	6 PM	7 PM	8 PM	CODE BULLETIN				
6 PM	7 PM	8 PM	9 PM	TELEPRINTER BULLETIN				
6⁴⁵ PM	7⁴⁵ PM	8⁴⁵ PM	9⁴⁵ PM	VOICE BULLETIN				
7 PM	8 PM	9 PM	10 PM	FAST CODE	SLOW CODE	FAST CODE	SLOW CODE	FAST CODE
8 PM	9 PM	10PM	11 PM	CODE BULLETIN				

W1AW's schedule is at the same local time throughout the year. The schedule according to your local time will change if your local time does not have seasonal adjustments that are made at the same time as North American time changes between standard time and daylight time. From the first Sunday in April to the last Sunday in October, UTC = Eastern Time + 4 hours. For the rest of the year, UTC = Eastern Time + 5 hours.

• **Morse code transmissions:**
Frequencies are 1.818, 3.5815, 7.0475, 14.0475, 18.0975, 21.0675, 28.0675 and 147.555 MHz.
Slow Code = practice sent at 5, $7^1/_2$, 10, 13 and 15 wpm.
Fast Code = practice sent at 35, 30, 25, 20, 15, 13 and 10 wpm.
Code practice text is from the pages of *QST*. The source is given at the beginning of each practice session and alternate speeds within each session. For example, "Text is from July 1992 *QST*, pages 9 and 81," indicates that the plain text is from the article on page 9 and mixed number/letter groups are from page 81. Code bulletins are sent at 18 wpm.
W1AW qualifying runs are sent on the same frequencies as the Morse code transmissions. West Coast qualifying runs are transmitted on approximately 3.590 MHz by K6YR. At the beginning of each code practice session, the schedule for the next qualifying run is presented. Underline one minute of the highest speed you copied, certify that your copy was made without aid, and send it to ARRL for grading. Please include your name, call sign (if any) and complete mailing address. Send a 9×12-inch SASE for a certificate, or a business-size SASE for an endorsement.

• **Teleprinter transmissions:**
Frequencies are 3.625, 7.095, 14.095, 18.1025, 21.095, 28.095 and 147.555 MHz.
Bulletins are sent at 45.45-baud Baudot and 100-baud AMTOR, FEC Mode B. 110-baud ASCII will be sent only as time allows.
On Tuesdays and Fridays at 6:30 PM Eastern Time, Keplerian elements for many amateur satellites are sent on the regular teleprinter frequencies.

• **Voice transmissions:**
Frequencies are 1.855, 3.99, 7.29, 14.29, 18.16, 21.39, 28.59 and 147.555 MHz.

• **Miscellanea:**
On Fridays, UTC, a DX bulletin replaces the regular bulletins.
W1AW is open to visitors from 10 AM until noon and from 1 PM until 3:45 PM on Monday through Friday. FCC licensed amateurs may operate the station during that time. Be sure to bring your current FCC amateur license or a photocopy.
In a communication emergency, monitor W1AW for special bulletins as follows: voice on the hour, teleprinter at 15 minutes past the hour, and CW on the half hour.
Headquarters and W1AW are closed on New Year's Day, President's Day, Good Friday, Memorial Day, Independence Day, Labor Day, Thanksgiving and the following Friday, and Christmas Day.

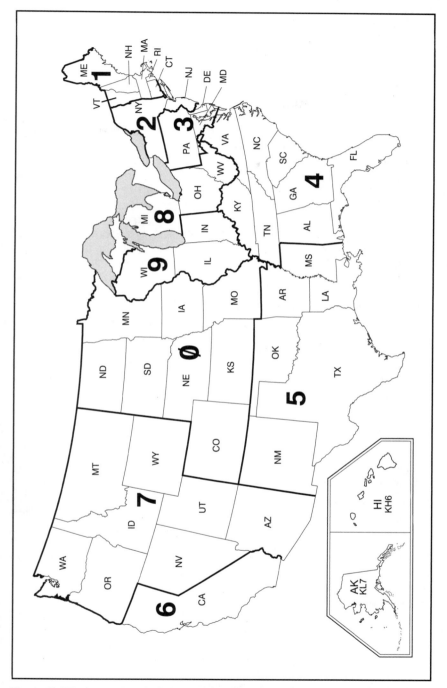

Figure 3—There are 10 US call areas. Hawaii is part of the sixth call area, and Alaska is part of the seventh.

Earning a License

Forms and Procedures

To renew or modify a license, you can file a copy of FCC Form 605. In addition, hams who have held a valid license that has expired within the past two years may apply for reinstatement with an FCC Form 605.

Licenses are normally good for ten years. Your application for a license renewal must be submitted to the FCC no more than 90 days before the license expires. (We recommend you submit the application for renewal between 90 and 60 days before your license expires.) If the FCC receives your renewal application before the license expires, you may continue to operate until your new license arrives, even if it is past the expiration date. If you forget to apply before your license expires, you may still be able to renew your license without taking another exam. There is a two-year grace period, during which you may apply for renewal of your expired license. Use an FCC Form 605 to apply for reinstatement (and your old call sign). If you apply for reinstatement of your expired license under this two-year grace period, you may not operate your station until your new license is issued. If you move or change addresses you should use an FCC Form 605 to notify the FCC of the change. If your license is lost or destroyed, however, just write a letter to the FCC explaining why you are requesting a new copy of your license.

You can ask one of the Volunteer Examiner Coordinators' offices to file your renewal application electronically if you don't want to mail the form to the FCC. You must still mail the form to the VEC, however. The ARRL/VEC Office will electronically file application forms. This service is free for any ARRL member.

Electronic Filing

You can also file your license renewal or address modification using the Universal Licensing System (ULS) on the World Wide Web. To use ULS, you must have an FCC Registration Number, or FRN. Obtain your FRN by registering with the Commission Registration System, known as CORES.

Described as an agency-wide registration system for anyone filing applications with or making payments to the FCC, CORES will assign a unique 10-digit FCC Registration Number, or FRN to all registrants. All Commission systems that handle financial, authorization of service, and enforcement activities will use the FRN. The FCC says use of the FRN will allow it to more rapidly verify fee payment. Amateurs mailing payments to the FCC — for example as part of a vanity call sign application — would include their FRN on FCC Form 159.

The on-line filing system and further information about CORES is available by visiting the FCC Web site, **www.fcc.gov** and clicking on the Commission Registration System link. Follow the directions on the Web site. It is also possible to register on CORES using a paper Form 160.

When you register with CORES you must supply a Taxpayer Identification Number, or TIN. For individuals, this is usually a Social

Security Number. Club stations must obtain an Assigned Taxpayer Identification Number (ATIN) before registering on CORES

Anyone can register on CORES and obtain an FRN. CORES/FRN is "entity registration." You don't need a license to be registered.

Once you have registered on CORES and obtained your FRN, you can proceed to renew or modify your license using the Universal Licensing System (ULS), also on the World Wide Web. Go to **www.fcc.gov/uls** and click on the "Online Filing" button. Follow the directions provided on the Web page to connect to the FCC's ULS database.

Paper Filing

The FCC has a set of detailed instructions for the Form 605, which are included with the form. To obtain a new Form 605, call the FCC Forms Distribution Center at 800-418-3676. You can also write to: Federal Communications Commission, Forms Distribution Center, 9300 E. Hampton Drive Capital Heights, MD 20743 (specify "Form 605" on the envelope). The Form 605 also is available from the FCC's fax on demand service. Call 202-418-0177 and ask for form number 000605. Form 605 also is available via the Internet. The World Wide Web location is: **www.fcc.gov/formpage.html** or you can receive the form via ftp to: **ftp.fcc.gov/pub/Forms/Form605**.

The ARRL/VEC has created a package that includes the portions of Form 605 that are needed for amateur applications, as well as a condensed set of instructions for completing the form. Write to: ARRL/VEC, Form 605, 225 Main Street, Newington, CT 06111-1494. (Please include a large business-sized stamped self-addressed envelope with your request.) **Figure 4** is a sample of those portions of an FCC Form 605 that you would complete to submit a change of address to the FCC.

Most of the form is simple to fill out. You will need to know that the Radio Service Code for box 1 is HA for Amateur Radio. (Just remember HAm radio.) You will have to include a "Taxpayer Identification Number" on the Form. This is normally your Social Security Number. If you don't want to write your Social Security Number on this form, then you can register with CORES as described above. When you receive your FRN from the FCC, you can use that number instead of your Social Security Number on the Form. Of course, you will have to supply your Social Security Number to register with the CORES.

The telephone number, fax number and e-mail address information is optional. The FCC will use that information to contact you in case there is a problem with your application.

Page two of the Form includes six General Certification Statements. Statement five may seem confusing. Basically, this statement means that you do not plan to install an antenna over 200 feet high, and that your permanent station location will not be in a designated wilderness area, wildlife preserve or nationally recognized scenic and recreational area.

The sixth statement indicates that you are familiar with the FCC RF Safety Rules, and that you will obey them. Chapter 10 (Subelement G0) of this book includes exam questions and explanations about the RF Safety Rules.

| FCC 605
Main Form | Quick-Form Application for Authorization in the Ship, Aircraft,
Amateur, Restricted and Commercial Operator,
and General Mobile Radio Services | Approved by OMB
3060 - 0850
See instructions for
public burden estimate |

1) Radio Service Code: H A

Application Purpose (Select only one) (   )

2) **NE** – New | **RO** – Renewal Only | **WD** – Withdrawal of Application
MD – Modification | **RM** – Renewal / Modification | **DU** – Duplicate License
AM – Amendment | **CA** – Cancellation of License | **AU** – Administrative Update

3)	If this request if for Developmental License or STA (Special Temporary Authorization) enter the appropriate code and attach the required exhibit as described in the instructions. Otherwise enter 'N' (Not Applicable).	(*N*) **D** **S** **N**/A ·
4)	If this request is for an Amendment or Withdrawal of Application, enter the file number of the pending application currently on file with the FCC.	File Number
5)	If this request is for a Modification, Renewal Only, Renewal / Modification, Cancellation of License, Duplicate License, or Administrative Update, enter the call sign (serial number for Commercial Operator) of the existing FCC license. If this is a request for consolidation of DO & DM Operator Licenses, enter serial number of DO.	Call Sign/Serial # W R 1 B
6)	If this request is for a New, Amendment, Renewal Only, or Renewal Modification, enter the requested expiration date of the authorization (this item is optional).	MM DD
7)	Does this filing request a Waiver of the Commission's rules? If 'Y', attach the required showing as described in the instructions.	(*N*) Yes No
8)	Are attachments (other than associated schedules) being filed with this application?	(*N*) Yes No

Applicant/Licensee Information

9) FCC Registration Number (FRN):   0 0 0 3 3 5 7 3 9 9

10) Applicant /Licensee is a(n): (*I*) Individual   Unicorporated Association   Trust   Government Entity   Joint Venture
Corporation   Limited Liability Corporation   Partnership   Consortium

11) First Name (if individual): *Larry*   MI: *D.*   Last Name: *Wolfgang*   Suffix:

11a) Date of Birth (required for Commercial Operators (including Restricted Radiotelephone)): **0 7** (mm)/ **0 9** (dd)/ **52** (yy)

12) Entity Name (if other than individual):

13) Attention To:

14) P.O. Box: And/Or   15) Street Address: *225 Main Street*

16) City: *Newington*   17) State: *CT*   18) Zip Code: *06111*   19) Country: *USA*

20) Telephone Number: *860-594-0200*   21) FAX Number:

22) E-Mail Address: *wr1b@arrl.net*

Figure 4—Portions of an FCC Form 605, showing the sections you would complete for a modification of your license, such as a change of address.

Ship Applicants/Licensees Only

23) Enter new name of vessel:_____

Fee Status

24) Is the applicant/licensee exempt from FCC application Fees? (N)Yes   No

25) Is the applicant/licensee exempt from FCC regulatory Fees? (N)Yes   No

General Certification Statements

1) The Applicant/Licensee waives any claim to the use of any particular frequency or of the electromagnetic spectrum as against the regulatory power of the United States because of the previous use of the same, whether by license or otherwise, and requests an authorization in accordance with this application.

2) The applicant/licensee certifies that all statements made in this application and in the exhibits, attachments, or documents incorporated by reference are material, are part of this application, and are true, complete, correct, and made in good faith.

3) Neither the Applicant/Licensee nor any member thereof is a foreign government or a representative thereof.

4) The applicant/licensee certifies that neither the applicant/licensee nor any other party to the application is subject to a denial of Federal benefits pursuant to Section 5301 of the Anti-Drug Abuse Act of 1988, 21 U.S.C. § 862, because of a conviction for possession or distribution of a controlled substance. **This certification does not apply to applications filed in services exempted under Section 1.2002© of the rules, 47 CFR § 1.2002©.** See Section 1.2002(b) of the rules, 47 CFR § 1.2002(b), for the definition of "party to the application" as used in this certification.

5) Amateur or GMRS Applicant/Licensee certifies that the construction of the station would NOT be an action which is likely to have a significant environmental effect (see the Commission's rules 47 CFR Sections 1.1301-1.1319 and Section 97.13(a) rules (available at web site http://wireless.fcc.gov/rules.html).

6) Amateur Applicant/Licensee certifies that they have READ and WILL COMPLY WITH Section 97.13(c) of the Commission's rules (available at web site http://wireless.fcc.gov/rules.html) regarding RADIOFREQUENCY (RF) RADIATION SAFETY and the amateur service section of OST/OET Bulletin Number 65 (available at web site http://www.fcc.gov/oet/info/documents/bulletins/).

Certification Statements For GMRS Applicants/Licensees

1) Applicant/Licensee certifies that he or she is claiming eligibility under Rule Section 95.5 of the Commission's rules.

2) Applicant/Licensee certifies that he or she is at least 18 years of age.

3) Applicant/Licensee certifies that he or she will comply with the requirement that use of frequencies 462.650, 467.650, 462.700 and 467.700 MHz is not permitted near the Canadian border North of Line A and East of Line C. These frequencies are used throughout Canada and harmful interference is anticipated.

4) Non-Individual applicants/licensees certify that they have NOT changed frequency or channel pairs, type of emission, antenna height, location of fixed transmitters, number of mobile units, area of mobile operation, or increase in power.

Signature

26) Typed or Printed Name of Party Authorized to Sign

First Name: *Larry*   MI: *D.*   Last Name: *Wolfgang*   Suffix:

27) Title:

Signature: *Larry D. Wolfgang*   28) Date: *4/14/2003*

Failure to Sign This Application May Result in Dismissal Of The Application And Forfeiture Of Any Fees Paid

WILLFUL FALSE STATEMENTS MADE ON THIS FORM OR ANY ATTACHMENTS ARE PUNISHABLE BY FINE AND/OR IMPRISONMENT (U.S. Code, Title 18, Section 1001) AND / OR REVOCATION OF ANY STATION LICENSE OR CONSTRUCTION PERMIT (U.S. Code, Title 47, Section 312(a)(1)), AND / OR FORFEITURE (U.S. Code, Title 47, Section 503).

FCC 605 – Main Form
October 2002 - Page 2

Volunteer Examiner Program

Before you can take an FCC exam, you'll have to fill out a copy of the National Conference of Volunteer Examiner Coordinators' (NCVEC) Quick Form 605. This form is used as an application for a new license or an upgraded license. The NCVEC Quick Form 605 is only used at license exam sessions. This form includes some information that the Volunteer Examiner Coordinator's office will need to process your application with the FCC. See **Figure 5**. You should not use an NCVEC Quick Form 605 to apply for a license renewal or modification with the FCC. *Never* mail these forms to the FCC, because that will result in a rejection of the application. Likewise, an FCC Form 605 can't be used for an exam application.

All US amateur exams are administered by Volunteer Examiners who are certified by a Volunteer-Examiner Coordinator (VEC). *The ARRL's FCC Rule Book* contains more details about the Volunteer-Examiner program.

To qualify for a Technician license you must pass Element 2. If you already hold a valid Novice license, then you have credit for passing Element 1 and you can earn a Technician license with Morse code credit. In that case you will be able to continue using your Novice HF privileges. In addition, your upgrade to Technician will earn you full Amateur privileges on the VHF, UHF and higher frequency bands.

The Element 2 exam consists of 35 questions taken from a pool of more than 350 questions. The question pools for all amateur exams are maintained by a Question Pool Committee selected by the Volunteer Examiner Coordinators. The FCC allows Volunteer Examiners to select the questions for an amateur exam, but they must use the questions exactly as they are released by the VEC that coordinates the test session. If you attend a test session coordinated by the ARRL/VEC, your test will be designed by the ARRL/VEC or by a computer program designed by the VEC. The questions and answers will be exactly as they are printed in this book.

Finding an Exam Opportunity

To determine where and when exams will be given in your area, contact the ARRL/VEC office, or watch for announcements in the Hamfest Calendar and Coming Conventions columns in *QST*. Many local clubs sponsor exams, so they are another good source of information on exam opportunities. Upcoming exams are listed on **ARRLWeb** at: **www.arrl.org/arrlvec/examsearch.phtml**. Registration deadlines, and the time and location of the exams, are mentioned prominently in publicity releases about upcoming sessions.

Taking the Exam

By the time examination day rolls around, you should have already prepared yourself. This means getting your schedule, supplies and mental attitude ready. Plan your schedule so you'll get to the examination site

NCVEC QUICK-FORM 605 APPLICATION FOR
AMATEUR OPERATOR/PRIMARY STATION LICENSE

SECTION 1 - TO BE COMPLETED BY APPLICANT

PRINT LAST NAME: Sayad   SUFFIX: FIRST NAME: Daniel   INITIAL: S   STATION CALL SIGN (IF ANY):

MAILING ADDRESS (Number and Street or P.O. Box): 225 Main St

SOCIAL SECURITY NUMBER / TIN (OR FCC LICENSEE ID #): 125-4236-54

CITY: Newington   STATE CODE: CT   ZIP CODE (5 or 9 Numbers): 06111   E-MAIL ADDRESS (OPTIONAL):

DAYTIME TELEPHONE NUMBER (Include Area Code) OPTIONAL: 860-594-0200   FAX NUMBER (Include Area Code) OPTIONAL:

ENTITY NAME (IF CLUB, MILITARY RECREATION, RACES):

Type of Applicant: ☒ Individual ☐ Amateur Club ☐ Military Recreation ☐ RACES (Modify Only)

CLUB, MILITARY RECREATION, OR RACES CALL SIGN

SIGNATURE OF RESPONSIBLE CLUB OFFICIAL

I HEREBY APPLY FOR (Make an X in the appropriate box(es))

☒ EXAMINATION for a **new** license grant

☐ EXAMINATION for **upgrade** of my license class

☐ CHANGE my **name** on my license to my new name

Former Name: _____
(Last name) (Suffix) (First name) (MI)

☐ CHANGE my mailing address to **above** address

☐ CHANGE my station **call sign** systematically

Applicant's Initials: _____

☐ RENEWAL of my license grant.

Do you have another license application on file with the FCC which has not been acted upon?   PURPOSE OF OTHER APPLICATION   PENDING FILE NUMBER (FOR VEC USE ONLY)

I certify that:
* I waive any claim to the use of any particular frequency regardless of prior use by license or otherwise;
* All statements and attachments are true, complete and correct to the best of my knowledge and belief and are made in good faith;
* I am not a representative of a foreign government;
* I am not subject to a denial of Federal benefits pursuant to Section 5301 of the Anti-Drug Abuse Act of 1988, 21 U.S.C. § 862;
* The construction of my station will NOT be an action which is likely to have a significant environmental effect (See 47 CFR Sections 1.301-1.319 and Section 97.13(a));
* I have read and WILL COMPLY with Section 97.13(c) of the Commission's Rules regarding RADIOFREQUENCY (RF) RADIATION SAFETY and the amateur service section of OST/OET Bulletin Number 65.

Signature of applicant (Do not print, type, or stamp. Must match applicant's name above.)

X _____   Date Signed: 1/16/03

SECTION 2 - TO BE COMPLETED BY ALL ADMINISTERING VEs

Applicant is qualified for operator license class:

☐ NO NEW LICENSE OR UPGRADE WAS EARNED

☒ TECHNICIAN   Element 2

☐ GENERAL   Elements 1, 2 and 3

☐ AMATEUR EXTRA   Elements 1, 2, 3 and 4

DATE OF EXAMINATION SESSION: 1/16/03

EXAMINATION SESSION LOCATION: Newington CT

VEC ORGANIZATION: ARRL

VEC RECEIPT DATE:

I CERTIFY THAT I HAVE COMPLIED WITH THE ADMINISTERING VE REQUIREMENTS IN PART 97 OF THE COMMISSION'S RULES AND WITH THE INSTRUCTIONS PROVIDED BY THE COORDINATING VEC AND THE FCC.

1st VEs NAME (Print First, MI, Last, Suffix)	VEs STATION CALL SIGN	VEs SIGNATURE (Must match name)	DATE SIGNED
DAVID C PATTON	NT1N		16 JAN 03
2nd VEs NAME (Print First, MI, Last, Suffix) PERRY T GREEN	WY1O		1/17/03
3rd VEs NAME (Print First, MI, Last, Suffix) Larry D. Wolfgang	WR1B	Larry D. Wolfgang	1/16/03

DO NOT SEND THIS FORM TO FCC – THIS IS NOT AN FCC FORM.
IF THIS FORM IS SENT TO FCC, FCC WILL RETURN IT TO YOU WITHOUT ACTION

NCVEC FORM 605 - FEBRUARY 2001
FOR VE/VEC USE ONLY - Page 1

Figure 5—An NCVEC Quick Form 605 as it would be completed for a new Technician license.

with plenty of time to spare. There's no harm in being early. In fact, you might have time to discuss hamming with another applicant, which is a great way to calm pretest nerves. Try not to discuss the material that will be on the examination, as this may make you even more nervous. By this time, it's too late to study anyway!

What supplies will you need? First, be sure you bring your current original Amateur Radio license, if you have one. Bring a photocopy of your license, too, as well as the original and a photocopy of any Certificates of Successful Completion of Examination (CSCE) that you plan to use for exam credit. Bring along several sharpened number 2 pencils and two pens (blue or black ink). Be sure to have a good eraser. A pocket calculator may also come in handy. You may use a programmable calculator if that is the kind you have, but take it into your exam "empty" (cleared of all programs and constants in memory). Don't program equations ahead of time, because you may be asked to demonstrate that there is nothing in the calculator memory. The examining team has the right to refuse a candidate the use of any calculator that they feel may contain information for the test or could otherwise be used to cheat on the exam.

The Volunteer Examiner Team is required to check two forms of identification before you enter the test room. This includes your *original* Amateur Radio license, if you have one—not a photocopy. A photo ID of some type is best for the second form of ID, but is not required by the FCC. Other acceptable forms of identification include a driver's license, a piece of mail addressed to you or a birth certificate.

The following description of the testing procedure applies to exams coordinated by the ARRL/VEC, although many other VECs use a similar procedure.

Code Test

The code test is usually given before the written exams. If you don't plan to take the code exam, just sit quietly while the other candidates give it a try.

Before you take the code test, you'll be handed a piece of paper to copy the code as it is sent. The test will begin with about a minute of practice copy. Then comes the actual test: at least five minutes of Morse code. You are responsible for knowing the 26 letters of the alphabet, the numerals 0 through 9, the period, comma, question mark, and the procedural signals $\overline{AR}$ (+), $\overline{SK}$, $\overline{BT}$ (= or double dash) and $\overline{DN}$ (/ or fraction bar, sometimes called the "slant bar").

You may copy the entire text word for word, or just take notes on the content. At the end of the transmission, the examiner will hand you 10 questions about the text. Fill in the blanks with your answers. (You must spell each answer exactly as it was sent.) If you get at least 7 correct, you pass! Alternatively, the exam team has the option to look at your copy sheet if you fail the 10-question exam. If you have one minute of solid copy (25 characters), the examiners can certify that you passed the test on that basis. The format of the test transmission is generally similar to one side of a normal on-the-air amateur conversation.

A sending test may not be required. The Commission has decided that if applicants can demonstrate receiving ability, they most likely can also send at that speed. But be prepared for a sending test, just in case! Subpart 97.503(a) of the FCC Rules says, "A telegraphy examination must be sufficient to prove that the examinee has the ability to send correctly by hand and to receive correctly by ear texts in the international Morse code at not less than the prescribed speed..."

Written Tests

After the code tests are administered, you'll take the written examination. The examiner will give each applicant a test booklet, an answer sheet and scratch paper. After that, you're on your own. The first thing to do is read the instructions. Be sure to sign your name every place it's called for. Do all of this at the beginning to get it out of the way.

Next, check the examination to see that all pages and questions are there. If not, report this to the examiner immediately. When filling in your answer sheet make sure your answers are marked next to the numbers that correspond to each question.

Go through the entire exam, and answer the easy questions first. Next, go back to the beginning and try the harder questions. Leave the really tough questions for last. Guessing can only help, as there is no additional penalty for answering incorrectly.

If you have to guess, do it intelligently: At first glance, you may find that you can eliminate one or more "distracters." Of the remaining responses, more than one may seem correct; only one is the best answer, however. To the applicant who is fully prepared, incorrect distracters to each question are obvious. Nothing beats preparation!

After you've finished, check the examination thoroughly. You may have read a question wrong or goofed in your arithmetic. Don't be overconfident. There's no rush, so take your time. Think, and check your answer sheet. When you feel you've done your best and can do no more, return the test booklet, answer sheet and scratch pad to the examiner.

The Volunteer-Examiner Team will grade the exam while you wait. The passing mark is 74%. (That means 26 out of 35 questions correct — or no more than 9 incorrect answers on the Element 2 exam.) You will receive a Certificate of Successful Completion of Examination (CSCE) showing all exam elements that you pass at that exam session. If you are already licensed, and you pass the exam elements required to earn a higher license class, the CSCE authorizes you to operate with your new privileges immediately. When you use these new privileges, you must sign your call sign followed by the slant mark ("/"; on voice, say "stroke" or "slant") and the letters "KT," if you are upgrading from a Novice to a Technician with code license. You only have to follow this special identification procedure until your new license is granted by the FCC, however.

If you pass only some of the exam elements required for a license, you will still receive a CSCE. That certificate shows what exam elements you passed, and is valid for 365 days. Use it as proof that you passed those exam elements so you won't have to take them over again next time you try for the license.

And Now, Let's Begin

The complete Technician question pool (Element 2) is printed in this book. Each chapter lists all the questions for a particular subelement (such as Control Operator Duties — T5). A brief explanation about the correct answer is given after each question.

Table 4 shows the study guide or syllabus for the Element 2 exam as released by the Volunteer-Examiner Coordinators' Question Pool Committee in December 2002. The syllabus lists the topics to be covered by the Technician exam, and so forms the basic outline for the remainder of this book. Use the syllabus to guide your study.

The question numbers used in the question pool refer to this syllabus. Each question number begins with a syllabus-point number (for example, T0C or T1E). The question numbers end with a two-digit number. For example, question T3B09 is the ninth question about the T3B syllabus point.

The Question Pool Committee designed the syllabus and question pool so there are the same number of points in each subelement as there are exam questions from that subelement. For example, two exam questions on the Technician exam must be from the "Radio Phenomena" subelement, so there are two groups for that point. These are numbered T3A and T3B. While not a requirement of the FCC Rules, the Question Pool Committee recommends that one question be taken from each group to make the best possible license exams.

Good luck with your studies!

Table 4
Technician Class (Element 2) Syllabus

(Required for all operator licenses.)

SUBELEMENT T1 — FCC Rules

[5 Exam Questions — 5 Groups]

T1A Definition and purpose of Amateur Radio Service, Amateur-Satellite Service in places where the FCC regulates these services and elsewhere; Part 97 and FCC regulation of the amateur services; Penalties for unlicensed operation and for violating FCC rules; Prohibited transmissions.

T1B International aspect of Amateur Radio; International and domestic spectrum allocation; Spectrum sharing; International communications; reciprocal operation.

T1C All about license grants; Station and operator license grant structure including responsibilities, basic differences; Privileges of the various operator license classes; License grant term; Modifying and renewing license grant; Grace period.

T1D Qualifying for a license; General eligibility; Purpose of examination; Examination elements; Upgrading operator license class; Element credit; Provision for physical disabilities.

T1E Amateur station call sign systems including Sequential, Vanity and Special Event; ITU Regions; Call sign formats.

SUBELEMENT T2 — Methods of Communication

[2 Exam Questions — 2 Groups]

T2A How Radio Works; Electromagnetic spectrum; Magnetic/Electric Fields; Nature of Radio Waves; Wavelength; Frequency; Velocity; AC Sine wave/Hertz; Audio and Radio frequency.

T2B Frequency privileges granted to Technician class operators; Amateur service bands; Emission types and designators; Modulation principles; AM/FM/Single sideband/upper-lower, international Morse code (CW), RTTY, packet radio and data emission types; Full quieting.

SUBELEMENT T3 — Radio Phenomena

[2 Exam Questions — 2 Groups]

T3A How a radio signal travels; Atmosphere/troposphere/ionosphere and ionized layers; Skip distance; Ground (surface)/sky (space) waves; Single/multihop; Path; Ionospheric absorption; Refraction.

T3B HF vs. VHF vs. UHF characteristics; Types of VHF-UHF propagation; Daylight and seasonal variations; Tropospheric ducting; Line of sight; Maximum usable frequency (MUF); Sunspots and sunspot Cycle, Characteristics of different bands.

SUBELEMENT T4 — Station Licensee Duties

[3 Exam Questions — 3 Groups]

T4A Correct name and mailing address on station license grant; Places from where station is authorized to transmit; Selecting station location; Antenna structure location; Stations installed aboard ship or aircraft.

T4B Designation of control operator; FCC presumption of control operator; Physical control of station apparatus; Control point; Immediate station control; Protecting against unauthorized transmissions; Station records; FCC Inspection; Restricted operation.

T4C Providing public service; emergency and disaster communications; Distress calling; Emergency drills and communications; Purpose of RACES.

SUBELEMENT T5 — Control Operator Duties

[3 Exam Questions — 3 Groups]

T5A Determining operating privileges, Where control operator must be situated while station is locally or remotely controlled; Operating other amateur stations.

T5B Transmitter power standards; Interference to stations providing emergency communications; Station identification requirements.

T5C Authorized transmissions, Prohibited practices; Third party communications; Retransmitting radio signals; One way communications.

SUBELEMENT T6 — Good Operating Practices

[3 Exam Questions — 3 Groups]

T6A Calling another station; Calling CQ; Typical amateur service radio contacts; Courtesy and respect for others; Popular Q-signals; Signal reception reports; Phonetic alphabet for voice operations.

T6B Occupied bandwidth for emission types; Mandated and voluntary band plans; CW operation.

T6C TVI and RFI reduction and elimination, Band/Low/High pass filter, Out of band harmonic Signals, Spurious Emissions, Telephone Interference, Shielding, Receiver Overload.

SUBELEMENT T7 — Basic Communications Electronics

[3 Exam Questions — 3 Groups]

T7A Fundamentals of electricity; AC/DC power; units and definitions of current, voltage, resistance, inductance, capacitance and impedance; Rectification; Ohm's Law principle (simple math); Decibel; Metric system and prefixes (e.g., pico, nano, micro, milli, deci, centi, kilo, mega, giga).

T7B Basic electric circuits; Analog vs. digital communications; Audio/RF signal; Amplification.

T7C Concepts of Resistance/resistor; Capacitor/capacitance; Inductor/Inductance; Conductor/Insulator; Diode; Transistor; Semiconductor devices; Electrical functions of and schematic symbols of resistors, switches, fuses, batteries, inductors, capacitors, antennas, grounds and polarity; Construction of variable and fixed inductors and capacitors.

SUBELEMENT T8 — Good Engineering Practice

[6 Exam Questions — 6 Groups]

T8A Basic amateur station apparatus; Choice of apparatus for desired communications; Setting up station; Constructing and modifying amateur station apparatus; Station layout for CW, SSB, FM, Packet and other popular modes.

T8B How transmitters work; Operation and tuning; VFO; Transceiver; Dummy load; Antenna switch; Power supply; Amplifier; Stability; Microphone gain; FM deviation; Block diagrams of typical stations.

T8C How receivers work, operation and tuning, including block diagrams; Superheterodyne including Intermediate frequency; Reception; Demodulation or Detection; Sensitivity; Selectivity; Frequency standards; Squelch and audio gain (volume) control.

T8D How antennas work; Radiation principles; Basic construction; Half wave dipole length vs. frequency; Polarization; Directivity; ERP;

Directional/non-directional antennas; Multiband antennas; Antenna gain; Resonant frequency; Loading coil; Electrical vs. physical length; Radiation pattern; Transmatch.

T8E How transmission lines work; Standing waves/SWR/SWR-meter; Impedance matching; Types of transmission lines; Feed point; Coaxial cable; Balun; Waterproofing Connections.

T8F Voltmeter/ammeter/ohmmeter/multi/S-meter, peak reading and RF watt meter; Building/modifying equipment; Soldering; Making measurements; Test instruments.

SUBELEMENT T9 — Special Operations

[2 Exam Questions — 2 Groups]

T9A How an FM Repeater Works; Repeater operating procedures; Available frequencies; Input/output frequency separation; Repeater ID requirements; Simplex operation; Coordination; Time out; Open/closed repeater; Responsibility for interference.

T9B Beacon, satellite, space, EME communications; Radio control of models; Autopatch; Slow scan television; Telecommand; CTCSS tone access; Duplex/crossband operation.

SUBELEMENT T0 — Electrical, Antenna Structure and RF Safety Practices

[6 Exam Questions — 6 Groups]

T0A Sources of electrical danger in amateur stations: lethal voltages, high current sources, fire; avoiding electrical shock; Station wiring; Wiring a three wire electrical plug; Need for main power switch; Safety interlock switch; Open/short circuit; Fuses; Station grounding.

T0B Lightning protection; Antenna structure installation safety; Tower climbing Safety; Safety belt/hard hat/safety glasses; Antenna structure limitations.

T0C Definition of RF radiation; Procedures for RF environmental safety; Definitions and guidelines.

T0D Radiofrequency exposure standards; Near/far field, Field strength; Compliance distance; Controlled/Uncontrolled environment.

T0E RF Biological effects and potential hazards; Radiation exposure limits; OET Bulletin 65; MPE (Maximum permissible exposure).

T0F Routine station evaluation.

FCC Rules

Your Technician exam (Element 2) will consist of 35 questions, taken from the Technician question pool, as prepared by the Volunteer Examiner Coordinators' Question Pool Committee. A certain number of questions are taken from each of the 10 subelements. There will be 5 questions from the FCC Rules subelement printed in this chapter. These questions are divided into 5 groups, labeled T1A through T1E

After most of the explanations in this chapter you will see a reference to Part 97 of the Federal Communications Commission Rules, set inside square brackets, like [97.3 (a) (5)]. This tells you where to look for the exact wording of the Rules as they relate to that question. For a complete copy of Part 97, along with simple explanations of the Rules governing Amateur Radio, see *The FCC Rule Book* published by the ARRL.

T1A   Definition and purpose of Amateur Radio Service, Amateur-Satellite Service in places where the FCC regulates these services and elsewhere; Part 97 and FCC regulation of the amateur services; Penalties for unlicensed operation and for violating FCC rules; Prohibited transmissions.

T1A01 Who makes and enforces the rules for the amateur service in the United States?
A. The Congress of the United States
B. The Federal Communications Commission (FCC)
C. The Volunteer Examiner Coordinators (VECs)
D. The Federal Bureau of Investigation (FBI)

B   Part 97 of the Federal Communication Commission's Rules governs the Amateur Radio Service in the United States. It is the FCC that enforces those rules.

T1A02 What are two of the five fundamental purposes for the amateur service in the United States?

 A. To protect historical radio data, and help the public understand radio history

 B. To help foreign countries improve communication and technical skills, and encourage visits from foreign hams

 C. To modernize radio schematic drawings, and increase the pool of electrical drafting people

 D. To increase the number of trained radio operators and electronics experts, and improve international goodwill

D In Section 97.1 of its Rules, the FCC describes the basis and purpose of the amateur service. It consists of five principles, which comprise the foundation and rationale for the existence of amateur radio in the US. Here's what the Rules say:

§97.1 Basis and purpose.

The rules and regulations in this Part are designed to provide an amateur radio service having a fundamental purpose as expressed in the following principles:

(a) Recognition and enhancement of the value of the amateur service to the public as a voluntary noncommercial communication service, particularly with respect to providing emergency communications.

(b) Continuation and extension of the amateur's proven ability to contribute to the advancement of the radio art.

(c) Encouragement and improvement of the amateur service through rules which provide for advancing skills in both the communications and technical phases of the art.

(d) Expansion of the existing reservoir within the amateur radio service of trained operators, technicians, and electronics experts.

(e) Continuation and extension of the amateur's unique ability to enhance international goodwill.

Although the question only requires that you identify the last two items on this list of five, you should be familiar with all of them. [97.1 (d) and (e)]

T1A03 What is the definition of an amateur station?

A. A radio station in a public radio service used for radiocommunications

B. A radio station using radiocommunications for a commercial purpose

C. A radio station using equipment for training new broadcast operators and technicians

D. A radio station in the amateur service used for radiocommunications

D   The FCC defines an amateur station as, "A station licensed in the amateur service, including the apparatus necessary for carrying on radio-communications." The person operating an amateur station has an interest in self-training, intercommunication and technical investigations or experiments. [97.3 (a) (5)]

T1A04 When is an amateur station authorized to transmit information to the general public?

A. Never

B. Only when the operator is being paid

C. Only when the broadcast transmission lasts less than 1 hour

D. Only when the broadcast transmission lasts longer than 15 minutes

A   Amateur Radio is a two-way communications service. Amateur Radio stations may not engage in broadcasting. Broadcasting normally means the transmission of information intended for reception by the general public. [97.113 (b)]

T1A05 When is an amateur station authorized to transmit music?

    A. Amateurs may not transmit music, except as an incidental part of an authorized rebroadcast of space shuttle communications

    B. Only when the music produces no spurious emissions

    C. Only when the music is used to jam an illegal transmission

    D. Only when the music is above 1280 MHz, and the music is a live performance

A   Under FCC Rules, amateurs may not transmit music of any form. This means you can't transmit your band's practice session or play the piano for transmission over the air. There is one exception to the "No Music" rule. If you obtain special permission from NASA to retransmit the audio from a space shuttle mission or the International Space Station for other amateurs to listen, and during that retransmission NASA or the astronauts play some music over the air, you won't get in trouble. [97.113 (a) (4), 97.113 (e)]

T1A06 When is the transmission of codes or ciphers allowed to hide the meaning of a message transmitted by an amateur station?

    A. Only during contests

    B. Only during nationally declared emergencies

    C. Codes and ciphers may not be used to obscure the meaning of a message, although there are special exceptions

    D. Only when frequencies above 1280 MHz are used

C   You can't use codes or ciphers to obscure the meaning of transmissions. This means you can't make up a "secret" code to send messages over the air to a friend. However, there are special exceptions. Control signals transmitted for remote control of model craft are not considered codes or ciphers. Neither are telemetry signals, such as a satellite might transmit to tell about its condition. A space station (satellite) control operator can use specially coded signals to control the satellite. [97.113 (a) (4), 97.211 (b), 97.217]

T1A07 Which of the following one-way communications may NOT be transmitted in the amateur service?

A. Telecommand to model craft
B. Broadcasts intended for reception by the general public
C. Brief transmissions to make adjustments to the station
D. Morse code practice

B   Amateur Radio is a two-way communications service. Nevertheless, there are exceptions that allow amateurs to make one-way transmissions to control model craft, make adjustments to their stations and even to transmit Morse code practice. However, Amateur Radio stations may not engage in broadcasting. Broadcasting normally means the transmission of information intended for reception by the general public. [97.3 (a) (10), 97.113 (b)]

T1A08 What is an amateur space station?

A. An amateur station operated on an unused frequency
B. An amateur station awaiting its new call letters from the FCC
C. An amateur station located more than 50 kilometers above the Earth's surface
D. An amateur station that communicates with the International Space Station

C   The FCC defines a space station as "An amateur station located more than 50 kilometers (km) — about 30 miles — above the Earth's surface." This obviously includes amateur satellites, and also includes any operation from the Space Shuttle, the International Space Station and any future operations by astronauts in space. [97.3 (a) (40)]

T1A09 Who may be the control operator of an amateur space station?

A. An amateur holding an Amateur Extra class operator license grant
B. Any licensed amateur operator
C. Anyone designated by the commander of the spacecraft
D. No one unless specifically authorized by the government

B   Any licensed ham can be the licensee or control operator of a space station. Likewise, any licensed amateur may operate through or communicate with a space station, as long as their transmissions take place on frequencies available for that license class. [97.207 (a)]

T1A10 When may false or deceptive signals or communications be transmitted by an amateur station?

    A. Never

    B. When operating a beacon transmitter in a "fox hunt" exercise

    C. When playing a harmless "practical joke"

    D. When you need to hide the meaning of a message for secrecy

A   Amateurs may not transmit false or deceptive signals, such as a distress call when no emergency exists. You must not, for example, start calling MAYDAY (an international distress signal) unless you are in a life-threatening situation. [97.113 (a) (4)]

T1A11 When may an amateur station transmit unidentified communications?

    A. Only during brief tests not meant as messages

    B. Only when they do not interfere with others

    C. Only when sent from a space station or to control a model craft

    D. Only during two-way or third-party communications

C   The rules prohibit unidentified communications or signals. These are signals where the transmitting station's call sign is not included. There are two exceptions to this rule, which exempt space stations and telecommand stations [97.119 (a)]

T1A12 What is an amateur communication called that does NOT have the required station identification?

    A. Unidentified communications or signals

    B. Reluctance modulation

    C. Test emission

    D. Tactical communication

A   Unidentified communications or signals are signals where the transmitting station's call sign is not included. Be sure you understand the proper station identification procedures, so you don't violate this rule. [97.119 (a)]

T1A13 What is a transmission called that disturbs other communications?

A. Interrupted CW
B. Harmful interference
C. Transponder signals
D. Unidentified transmissions

B   A transmission that disturbs other authorized communications is called harmful interference. FCC Rules define harmful interference as, "Interference which endangers the functioning of a radionavigation service or of other safety services or seriously degrades, obstructs or repeatedly interrupts a radio-communication service operating in accordance with the Radio Regulations. [97.3 (a) (23)]

T1A14 What does the term broadcasting mean?

A. Transmissions intended for reception by the general public, either direct or relayed
B. Retransmission by automatic means of programs or signals from non-amateur stations
C. One-way radio communications, regardless of purpose or content
D. One-way or two-way radio communications between two or more stations

A   Amateur Radio stations may not engage in broadcasting. Broadcasting normally means the transmission of information intended for reception by the general public. These broadcast transmissions may either be direct or relayed. [97.3 (a) (10)]

T1A15 Why is indecent and obscene language prohibited in the Amateur Service?

A. Because it is offensive to some individuals
B. Because young children may intercept amateur communications with readily available receiving equipment
C. Because such language is specifically prohibited by FCC Rules
D. All of these choices are correct

D   Amateurs may not use obscene or indecent language. It is prohibited by the Rules. You should also remember that anyone, of any age, can hear your transmission if they happen to be tuned to your transmitting frequency. Depending on your operating frequency and other conditions, your signals can be heard around the world. While there is no list of "banned words" or other specific list, you should avoid any questionable language. [97.113 (a) (4)]

T1A16 Which of the following is a prohibited amateur radio transmission?

    A. Using an autopatch to seek emergency assistance
    B. Using an autopatch to pick up business messages
    C. Using an autopatch to call for a tow truck
    D. Using an autopatch to call home to say you are running late

B   FCC Rules prohibit amateur operators from engaging in any type of business communications. You may not conduct communications for your own business, or for your employer. This includes using the autopatch on your local repeater. [97.113 (a) (3)]

T1B International aspect of Amateur Radio; International and domestic spectrum allocation; Spectrum sharing; International communications; reciprocal operation; International and domestic spectrum allocation; Spectrum sharing; International communications; reciprocal operation.

T1B01 What are the frequency limits of the 6-meter band in ITU Region 2?

    A. 52.0 - 54.5 MHz
    B. 50.0 - 54.0 MHz
    C. 50.1 - 52.1 MHz
    D. 50.0 - 56.0 MHz

B   ITU Region 2 comprises North and South America as well as the Caribbean Islands and Hawaii. Alaska is in ITU Region 2. In those areas of Region 2 that are under FCC Rules, the 6-meter band extends from 50.0 to 54.0 MHz. [97.301 (a)]

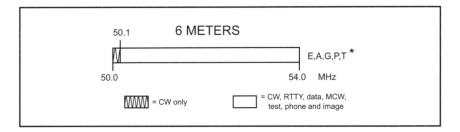

T1B02 What are the frequency limits of the 2-meter band in ITU Region 2?

    A. 144.0 - 148.0 MHz
    B. 145.0 - 149.5 MHz
    C. 144.1 - 146.5 MHz
    D. 144.0 - 146.0 MHz

A   The 2-meter band in Region 2 extends from 144.0 to 148.0 MHz. [97.301 (a)]

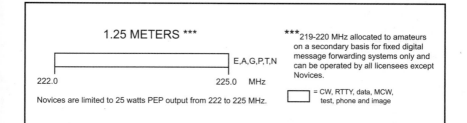

T1B03 What are the frequency limits of the 1.25-meter band in ITU Region 2?

    A. 225.0 - 230.5 MHz
    B. 222.0 - 225.0 MHz
    C. 224.1 - 225.1 MHz
    D. 220.0 - 226.0 MHz

B   The 1.25-meter band in ITU Region 2 extends from 222.0 to 225.0 MHz. [97.301 (f)]

T1B04 What are the frequency limits of the 70-centimeter band in ITU Region 2?

    A. 430.0 - 440.0 MHz
    B. 430.0 - 450.0 MHz
    C. 420.0 - 450.0 MHz
    D. 432.0 - 435.0 MHz

C   The 70-centimeter band in ITU Region 2 extends from 420.0 to 450.0 MHz. [97.301 (a)]

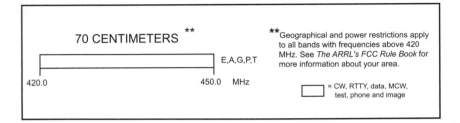

T1B05 What are the frequency limits of the 33-centimeter band in ITU Region 2?

    A. 903 - 927 MHz
    B. 905 - 925 MHz
    C. 900 - 930 MHz
    D. 902 - 928 MHz

D   The 33-centimeter band in ITU Region 2 extends from 902 to 928 MHz. [97.301 (a)]

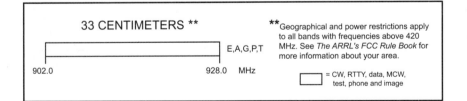

T1B06 What are the frequency limits of the 23-centimeter band in ITU Region 2?

- A. 1260 - 1270 MHz
- B. 1240 - 1300 MHz
- C. 1270 - 1295 MHz
- D. 1240 - 1246 MHz

B The 23-centimeter band in ITU Region 2 extends from 1240 to 1300 MHz. [97.301 (a)]

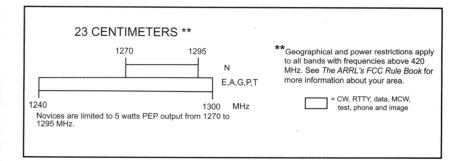

T1B07 What are the frequency limits of the 13-centimeter band in ITU Region 2?

- A. 2300 - 2310 MHz and 2390 - 2450 MHz
- B. 2300 - 2350 MHz and 2400 - 2450 MHz
- C. 2350 - 2380 MHz and 2390 - 2450 MHz
- D. 2300 - 2350 MHz and 2380 - 2450 MHz

A The 13-centimeter band in ITU Region 2 extends from 2300 to 2310 MHz and 2390 to 2450 MHz. [97.301 (a)]

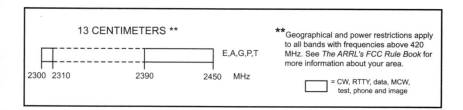

T1B08 If the FCC rules say that the amateur service is a secondary user of a frequency band, and another service is a primary user, what does this mean?

    A. Nothing special; all users of a frequency band have equal rights to operate

    B. Amateurs are only allowed to use the frequency band during emergencies

    C. Amateurs are allowed to use the frequency band only if they do not cause harmful interference to primary users

    D. Amateurs must increase transmitter power to overcome any interference caused by primary users

    C   A radio service that is designated as the primary service on a band is protected from interference caused by other radio services. A radio service that is designated as the secondary service must not cause harmful interference to, and must accept interference from, stations in a primary service. The amateur service has many different frequency bands. Some of them are allocated on a primary basis and some are secondary. [97.303]

T1B09 What rule applies if two amateur stations want to use the same frequency?

    A. The station operator with a lesser class of license must yield the frequency to a higher-class licensee

    B. The station operator with a lower power output must yield the frequency to the station with a higher power output

    C. Both station operators have an equal right to operate on the frequency

    D. Station operators in ITU Regions 1 and 3 must yield the frequency to stations in ITU Region 2

    C   If we are to make the best use of the limited amount of available spectrum, there must be ways to ensure that harmful interference is kept to a minimum. The FCC Rules require amateurs to cooperate with each other. If two amateur stations want to use the same frequency, both stations have an equal right to do so. [97.101 (b)]

T1B10 If you are operating on 28.400 MHz, in what amateur band are you operating?

A. 80 meters
B. 40 meters
C. 15 meters
D. 10 meters

D   Technician class amateurs who have passed the 5 word-per-minute Morse code exam may operate on 28.400 MHz. When they do, they are operating on the 10-meter band. [97.301 (e)]

T1B11 If you are operating on 223.50 MHz, in what amateur band are you operating?

A. 15 meters
B. 10 meters
C. 2 meters
D. 1.25 meters

D   If you are transmitting on 223.50 MHz, you are operating on the 1.25-meter band. See the chart with question T1B03. [97.301 (f)]

T1B12 When are you allowed to communicate with an amateur in a foreign country?

A. Only when the foreign amateur uses English
B. Only when you have permission from the FCC
C. Only when a third party agreement exists between the US and the foreign country
D. At any time, unless it is not allowed by either government

D   You may converse with any amateur station at any time. This includes amateurs in foreign countries, unless either amateur's government prohibits the communications. (There are times when a government will not allow its amateurs to talk with people in other countries.) [97.111 (a) (1)]

T1B13 If you are operating FM phone on the 23-cm band and learn that you are interfering with a radiolocation station outside the US, what must you do?

A. Stop operating or take steps to eliminate this harmful interference

B. Nothing, because this band is allocated exclusively to the amateur service

C. Establish contact with the radiolocation station and ask them to change frequency

D. Change to CW mode, because this would not likely cause interference

A   In the entire 23-cm band (1240 - 1300 MHz) no amateur station shall cause harmful interference to, nor is protected from interference due to the operation of, stations authorized by other nations in the radiolocation service. In other words, if you learn that you are interfering with a foreign radiolocation station, you must stop operating or take steps to eliminate this harmful interference. [97.303 (h)]

T1B14 What does it mean for an amateur station to operate under reciprocal operating authority?

A. The amateur is operating in a country other than his home country

B. The amateur is allowing a third party to talk to an amateur in another country

C. The amateur has permission to communicate in a foreign language

D. The amateur has permission to communicate with amateurs in another country

A   It is interesting to note that many countries make arrangements for Amateurs from other countries to operate Amateur Radio while they are visiting. This is called reciprocal operating authority. [97.107]

T1B15 What are the frequency limits for the amateur radio service for stations located north of Line A in the 70-cm band?

- A. 430 - 450 MHz
- B. 420 - 450 MHz
- C. 432 - 450 MHz
- D. 440 - 450 MHz

A　　If you want to operate in the 420 to 430 MHz segment of the 70-centimeter band you should be aware of regulatory limitations. If you live within about 50 miles of the Canadian border you may be north of "Line A." The sharing requirements of Section 97.303(f)(1) say "No amateur station shall transmit from north of Line A in the 420 - 430 MHz segment." That means if you are operating from a location north of Line A, you are limited to the 430 to 450 MHz section of the 70-centimeter band. [97.301 (f) (1)]

T1C　　**All about license grants; Station and operator license grant structure including responsibilities, basic differences; Privileges of the various operator license classes; License grant term; Modifying and renewing license grant; Grace period.**

T1C01 Which of the following is required before you can operate an amateur station in the US?

- A. You must hold an FCC operator's training permit for a licensed radio station
- B. You must submit an FCC Form 605 together with a license examination fee
- C. The FCC must grant you an amateur operator/primary station license
- D. The FCC must issue you a Certificate of Successful Completion of Amateur Training

C　　An Amateur Radio license is really two licenses in one — an operator license and a station license. The operator license is one that lets you operate a station within your authorized privileges on amateur-service frequencies. You must have an amateur license to operate a transmitter on amateur service frequencies. The station license authorizes you to have an amateur station and its associated equipment. It also lists the call sign that identifies that station. The FCC calls this license an amateur operator/primary station license. One piece of paper includes both the operator and the station license. [97.5 (a)]

T1C02 **What are the US amateur operator licenses that a new amateur might earn?**

A. Novice, Technician, General, Advanced
B. Technician, Technician Plus, General, Advanced
C. Novice, Technician Plus, General, Advanced
D. Technician, Technician with Morse code, General, Amateur Extra

D   Anyone earning a new Amateur Radio license can earn one of three license classes — Technician, General and Amateur Extra. These vary in degree of knowledge required and frequency privileges granted. Technicians who have passed a Morse code exam have additional privileges at HF. [97.9 (a)]

T1C03 **How soon after you pass the examination elements required for your first Amateur Radio license may you transmit?**

A. Immediately
B. 30 days after the test date
C. As soon as the FCC grants you a license and the data appears in the FCC's ULS data base
D. As soon as you receive your license from the FCC

C   Your Amateur Radio license is valid as soon as the FCC grants the license and posts the information about your license in their electronic database. You don't have to wait for the actual license document to arrive in the mail before you begin to transmit. You can check the FCC database on one of the Internet license "servers" (or someone else can check it for you). There is a call sign lookup service on **ARRLWeb: www.arrl.org/fcc/fcclook.php3**. [97.5 (a)]

T1C04 **How soon before the expiration date of your license may you send the FCC a completed Form 605 or file with the Universal Licensing System on the World Wide Web for a renewal?**

A. No more than 90 days
B. No more than 30 days
C. Within 6 to 9 months
D. Within 6 months to a year

A   About 60 to 90 days before your present license expires (90 days maximum), you should apply for your license renewal. Use Form 605 or file with the Universal Licensing System on the World Wide Web for license renewal. [97.21 (a) (3) (i)]

T1C05 What is the normal term for an amateur station license grant?

A. 5 years
B. 7 years
C. 10 years
D. For the lifetime of the licensee

C   The FCC issues all licenses for a 10-year term. You should always renew your license for another 10 years before it expires. [97.25 (a)]

T1C06 What is the "grace period" during which the FCC will renew an expired 10-year license?

A. 2 years
B. 5 years
C. 10 years
D. There is no grace period

A   If you do forget to renew your license, you have up to two years to apply for renewal. After the two-year grace period, you will have to take the exam again. Your license is not valid during this two-year grace period, however. You may not operate an amateur station with an expired license. All the grace period means is that the FCC will renew the license if you apply during that time. [97.21 (b)]

T1C07 What is your responsibility as a station licensee?

A. You must allow another amateur to operate your station upon request
B. You must be present whenever the station is operated
C. You must notify the FCC if another amateur acts as the control operator
D. You are responsible for the proper operation of the station in accordance with the FCC rules

D   Part 97 of the FCC Rules provides quite a bit of guidance about what you can and cannot do as an Amateur Radio operator. Of course every possible situation can't be covered, so you will have to use good judgement in your operating practices. You should be familiar with the specific requirements of the Rules, and always try to operate within the intent of those Rules. As an Amateur Radio operator, you are always responsible for the proper operation of your Amateur Radio station. [97.103 (a)]

T1C08 Where does a US amateur license allow you to operate?

A. Anywhere in the world
B. Wherever the amateur service is regulated by the FCC
C. Within 50 km of your primary station location
D. Only at the mailing address printed on your license

B   A US amateur license allows you to operate wherever the FCC regulates the amateur service. That includes the 50 states as well as any territories under US government control. [97.5 (d)]

T1C09 Under what conditions are amateur stations allowed to communicate with stations operating in other radio services?

A. Never; amateur stations are only permitted to communicate with other amateur stations
B. When authorized by the FCC or in an emergency
C. When communicating with stations in the Citizens Radio Service
D. When a commercial broadcast station is using Amateur Radio frequencies for newsgathering during a natural disaster

B   You are allowed to communicate with stations in other radio services when authorized by the FCC or during a communications emergency. Section 97.403 of the Rules defines a communications emergency as "communication needs in connection with the immediate safety of human life and immediate protection of property when normal communications systems are not available." [97.111 (a) (3), (4) and 97.111 (b) (4).]

T1C10 To what distance limit may Technician class licensees communicate?

A. Up to 200 miles
B. There is no distance limit
C. Only to line of sight contacts distances
D. Only to contacts inside the USA

B   There is no FCC Rule that limit the distance over which you can communicate by Amateur Radio. The world is literally at your fingertips! The VHF and UHF bands available to a Technician licensee are often considered to be local communications bands, but there are ways to communicate over great distances on these bands.

T1C11 If you forget to renew your amateur license and it expires, may you continue to transmit?

A. No, transmitting is not allowed

B. Yes, but only if you identify using the suffix "GP"

C. Yes, but only during authorized nets

D. Yes, any time for up to two years (the "grace period" for renewal)

A   If you do forget to renew your license, you have up to two years to apply for a new license. After the two-year grace period, you will have to take the exam again. Your license is not valid during this two-year grace period, however. You may not operate an amateur station with an expired license.

T1D   **Qualifying for a license; General eligibility; Purpose of examination; Examination elements; Upgrading operator license class; Element credit; Provision for physical disabilities.**

T1D01 Who can become an amateur licensee in the US?

A. Anyone except a representative of a foreign government

B. Only a citizen of the United States

C. Anyone except an employee of the US government

D. Anyone

A   Amateur Radio is open to (almost) everyone. Anyone, except an agent or representative of a foreign government, is eligible to qualify for an Amateur Radio operator license. [97.5 (b) (1)]

T1D02 What age must you be to hold an amateur license?

A. 14 years or older

B. 18 years or older

C. 70 years or younger

D. There are no age limits

D   There is no age requirement for individuals to hold an Amateur Radio operator license. You can't be too young — or too old — to qualify for a license. [97.5 (b) (1)]

T1D03 What government agency grants your amateur radio license?

 A. The Department of Defense
 B. The State Licensing Bureau
 C. The Department of Commerce
 D. The Federal Communications Commission

 D In the United States, the Federal Communications Commission issues amateur licenses. A US amateur license allows you to operate wherever the FCC regulates the amateur service.

T1D04 What element credit is earned by passing the Technician class written examination?

 A. Element 1
 B. Element 2
 C. Element 3
 D. Element 4

 B The FCC refers to the various exams for Amateur Radio licenses as exam Elements. For example, exam Element 1 is the 5 word-per-minute (wpm) Morse code exam. Element 2 is the Technician written exam. [97.501 (c)]

T1D05 If you are a Technician licensee who has passed a Morse code exam, what is one document you can use to prove that you are authorized to use certain amateur frequencies below 30 MHz?

 A. A certificate from the FCC showing that you have notified them that you will be using the HF bands
 B. A certificate showing that you have attended a class in HF communications
 C. A Certificate of Successful Completion of Examination showing that you have passed a Morse code exam
 D. No special proof is required

 C When you pass the Morse code exam you will receive a Certificate of Successful Completion of Examination (CSCE) from the Volunteer Examiners who conduct the exam session. The CSCE proves that you passed the Morse code exam, and allows you to operate on the four HF Novice subbands that normally provide direct worldwide communication. [97.9 (b)]

T1D06 What is a Volunteer Examiner (VE)?

A. A certified instructor who volunteers to examine amateur teaching manuals

B. An FCC employee who accredits volunteers to administer amateur license exams

C. An amateur, accredited by one or more VECs, who volunteers to administer amateur license exams

D. An amateur, registered with the Electronic Industries Association, who volunteers to examine amateur station equipment

C   Volunteer Examiner Coordinators (VECs) maintain the question pools for all amateur exams. A VEC is an organization that has made an agreement with the FCC to coordinate Amateur Radio license examinations by using Volunteer Examiners (VEs). A VE is a licensed Amateur Radio operator who volunteers to help administer amateur license exams. [97.509 (a)]

T1D07 What minimum examinations must you pass for a Technician amateur license?

A. A written exam, Element 1 and a 5 WPM code exam, Element 2

B. A 5 WPM code exam, Element 1 and a written exam, Element 3

C. A single 35 question multiple choice written exam, Element 2

D. A written exam, Element 2 and a 5 WPM code exam, Element 4

C   The Technician class exam consists of a 35-question written test, Element 2. This book contains every question in the question pool from which your exam will be composed. If you study this book carefully, you will easily pass the Technician class exam. [97.503 (b) (1)]

T1D08 How may an Element 1 exam be administered to an applicant with a physical disability?

A. It may be skipped if a doctor signs a statement saying the applicant is too disabled to pass the exam

B. By holding an open book exam

C. By lowering the exam's pass rate to 50 percent correct

D. By using a vibrating surface or flashing light

D   The administering Volunteer Examiners must accommodate an examinee whose physical disabilities require a special examination procedure. (The administering VEs may require a physician's certification indicating the nature of the disability before determining which, if any, special procedures must be used.) An example of a special procedure for Element 1 would be to send the candidate Morse code using a vibrating surface or flashing light. [97.509 (k) and VE Instructions]

T1D09 What is the purpose of the Element 1 examination?

A. To test Morse code comprehension at 5 words-per-minute
B. To test knowledge of block diagrams
C. To test antenna-building skills
D. To test knowledge of rules and regulations

A   The FCC refers to the various exams for Amateur Radio licenses as exam Elements. For example, exam Element 1 is the 5 word-per-minute (wpm) Morse code exam. The purpose of Element 1 is to test Morse code comprehension at 5 words-per-minute. [97.503 (a)]

T1D10 If a Technician class licensee passes only the 5 words-per-minute Morse code test at an exam session, how long will this credit be valid for license upgrade purposes?

A. 365 days
B. Until the current license expires
C. Indefinitely
D. Until two years following the expiration of the current license

A   When you pass the Morse code exam, you will be issued a Certificate of Successful Completion of Examination (CSCE) for Element 1. Make sure you keep this paper in a safe place. Along with your actual Technician license, it serves as authorization to operate on the HF "Novice" bands. That proof is good for as long as you hold your Technician class license.

You can also use your CSCE as proof that you passed the Morse code exam for credit to complete an upgrade to a General class license. That credit is only valid for 365 days from the date on the form. After that you will have to pass the code exam again to upgrade to General. [97.505 (a) (6)]

T1D11 This question has been withdrawn.

T1E Amateur station call sign systems including Sequential, Vanity and Special Event; ITU Regions; Call sign formats.

T1E01 Which of the following call signs is a valid US amateur call?
A. UZ4FWD
B. KBL7766
C. KB3TMJ
D. VE3BKJ

C   The first letter of a US call will always be A, K, N or W. These letters are assigned to the United States as amateur call-sign prefixes. After the first one or two letters there will be a single number and then one to three letters. The letters before the number make up the call sign prefix, and the letters after the number are the suffix. Other countries use different prefixes — LA2UA is a Norwegian call sign, VE3BKJ is a call sign from Canada and VU2HO is from India.

T1E02 What letters must be used for the first letter in US amateur call signs?

A. K, N, U and W
B. A, K, N and W
C. A, B, C and D
D. A, N, V and W

B   The first letter of a US call will always be A, K, N or W. These letters are assigned to the United States as amateur call-sign prefixes. If the first letter is an A, then there will always be a second letter that follows that A. The second letter will be taken from the block that begins with A and ends with L. In other words, this block comprises AA through AL.

T1E03 What numbers are normally used in US amateur call signs?

A. Any two-digit number, 10 through 99
B. Any two-digit number, 22 through 45
C. A single digit, 1 though 9
D. A single digit, 0 through 9

D   The number in a US call sign shows the district where the call was first issued. Every US amateur call sign includes a single-digit number, 0 through 9 corresponding to these call districts. Amateurs may keep their calls when they move from one district to another. This means the number is not always an indication of where an amateur is located.

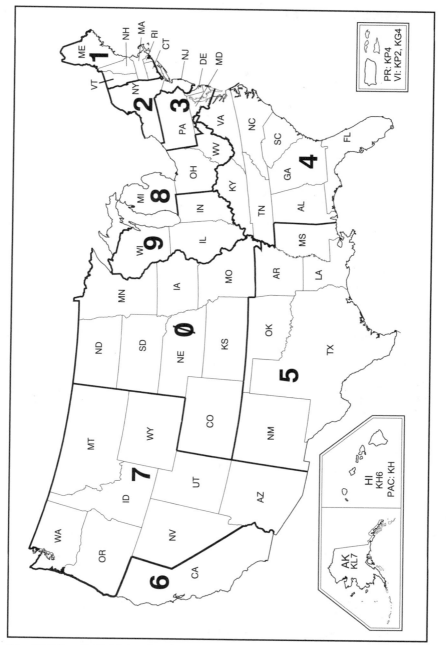

The 10 US call districts. An amateur holding the call sign K1STO lived in the first district when the FCC assigned her that call. Alaska is part of the seventh call district, but has its own set of prefixes: AL7, KL7, NL7 and WL7. Hawaii, part of the sixth district, has the AH6, KH6, NH6 and WH6 prefixes.

T1E04 In which ITU region is Alaska?

- A. ITU Region 1
- B. ITU Region 2
- C. ITU Region 3
- D. ITU Region 4

B   ITU Region 1 comprises Africa, Europe, Russia and parts of the Middle East. ITU Region 2 comprises North and South America as well as the Caribbean Islands and Hawaii. Alaska is in ITU Region 2.

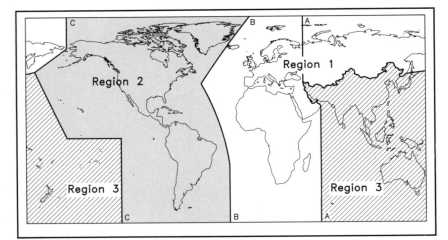

This map shows the world divided into three International Telecommunication Union (ITU) Regions.

T1E05 In which ITU region is Guam?

- A. ITU Region 1
- B. ITU Region 2
- C. ITU Region 3
- D. ITU Region 4

C   ITU Region 3 comprises Australia, China, India and parts of the Middle East. Region 3 also includes many islands in the Pacific Ocean, such as American Samoa, the Northern Mariana Islands, Guam and Wake Island.

T1E06 What must you transmit to identify your amateur station?

A. Your "handle"
B. Your call sign
C. Your first name and your location
D. Your full name

B   The FCC Rules require that you transmit your call sign to identify your station. You only have to ID at the end of a QSO and at least once every 10 minutes during its course. There is no legal requirement to transmit your call sign at the beginning of a contact. [97.119 (a)]

T1E07 How might you obtain a call sign made up of your initials?

A. Under the vanity call sign program
B. In a sequential call sign program
C. In the special event call sign program
D. There is no provision for choosing a call sign

A   In addition to the automatic sequential call sign system, the FCC has a way for Amateurs to select their own call sign. This is called the Vanity Call Sign program. Vanity call sign choices must take the same format as a sequential call sign for your license class. As a Technician licensee, you would be eligible to select a "one-by-three" or a "two-by-three" format call sign of your choice. You must select a call sign that is not currently assigned to another station. Call signs with the operator's initials for the suffix are popular. [97.19]

T1E08 How may an amateur radio licensee change his call sign without applying for a vanity call?

A. By requesting a systematic call sign change on an NCVEC Form 605
B. Paying a Volunteer Examiner team to process a call sign change request
C. By requesting a specific new call sign on an NCVEC Form 605 and sending it to the FCC in Gettysburg, PA
D. Contacting the FCC ULS database using the Internet to request a call sign change

A   If you would like to change your call sign, but don't want to pay the vanity fee, you can apply for a different call sign from the sequential call sign system. One simple way to do this is to complete an NCVEC Quick Form 605 and check the box to "Change my station call sign systematically." Initial the line next to this box and submit the application to a Volunteer Examiner Coordinator to file for you. The VEC may charge a fee for this service. The ARRL/VEC will file your application electronically for ARRL members for free. [97.21 (a) (3) (ii)]

T1E09 How may an amateur radio club obtain a station call sign?

A. You must apply directly to the FCC in Gettysburg, PA
B. You must apply through a Club Station Call Sign Administrator
C. You must submit FCC Form 605 to FCC in Washington, DC
D. You must notify VE team on NCVEC Form 605

B   Amateur Radio clubs can also obtain a station call sign. The club must name one member as the license trustee, to have primary responsibility for the license. The club must apply for the license through an FCC-approved Club Station Call Sign Administrator. The Call Sign Administrator collects the required information and sends it to the FCC in an electronic file. The ARRL/VEC, W4VEC and W5YI-VEC are Club Station Call Sign Administrators. [97.17 (b) (2)]

T1E10 Amateurs of which license classes are eligible to apply for temporary use of a 1-by-1 format Special Event call sign?

A. Only Amateur Extra class amateurs
B. 1-by-1 format call signs are not authorized in the US Amateur Service
C. Any FCC-licensed amateur
D. Only trustees of amateur radio clubs

C   Individuals or club groups who plan to operate an amateur station to commemorate a special event can apply for a Special-Event Call Sign. These special "one-by-one" format call signs can help call attention to the on-the-air operation at the special event. These call signs are issued for a short-term operation, normally 15 days or less. Any licensed amateur is allowed to apply for a Special-Event Call Sign. The FCC-approved Special-Event Call Sign Administrators coordinate these call signs. The ARRL/VEC, Laurel (Maryland) Amateur Radio Club, W4VEC and W5YI-VEC are Special-Event Call Sign Administrators.

T1E11 How does the FCC issue new amateur radio call signs?

A. By call sign district in random order
B. The applicant chooses a call sign no one else is using
C. By ITU prefix letter(s), call sign district numeral and a suffix in strict alphabetic order
D. The Volunteer Examiners who gave the exams choose a call sign no one else is using

C   The FCC issues call signs on a systematic basis. When they process your application, you get the next call sign to come out of the computer. This is done by ITU prefix letter(s), call sign district numeral and suffix in strict alphabetic order. [97.17 (d)]

T1E12 Which station call sign format groups are available to Technician Class amateur radio operators?

A. Group A
B. Group B
C. Only Group C
D. Group C and D

D   US Amateur call signs can have several different formats. Technician and General class call signs might have a "one-by-three" format — a letter followed by a number and then three more letters. These are called "Group C" call signs. They may also have a "two-by-three" format. These are called "Group D" call signs, which were originally intended for Novice licenses and club station call signs. (The FCC is not issuing any new Novice licenses.)

Methods of Communication

There will be 2 questions on your exam taken from the Methods of Communication subelement printed in this chapter. These questions are divided into 2 groups, labeled T2A and T2B

After some of the explanations in this chapter you will see a reference to Part 97 of the Federal Communications Commission Rules, set inside square brackets, like [97.3a5]. This tells you where to look for the exact wording of the Rules as they relate to that question. For a complete copy of Part 97, along with simple explanations of the Rules governing Amateur Radio, see *The ARRL's FCC Rule Book*.

T2A   How Radio Works; Electromagnetic spectrum; Magnetic/Electric Fields; Nature of Radio Waves; Wavelength; Frequency; Velocity; AC Sine wave/Hertz; Audio and Radio frequency.

T2A01 What happens to a signal's wavelength as its frequency increases?

A. It gets shorter
B. It gets longer
C. It stays the same
D. It disappears

A   A signal's wavelength is the distance that it travels during one complete cycle. As frequency increases, the time required to complete one cycle goes down by a proportionate amount. That also means that as the frequency goes up the wavelength (distance traveled during one cycle) goes down. If the frequency is doubled, the wavelength is cut in half.

T2A02 How does the frequency of a harmonic compare to the desired transmitting frequency?
- A. It is slightly more than the desired frequency
- B. It is slightly less than the desired frequency
- C. It is exactly two, or three, or more times the desired frequency
- D. It is much less than the desired frequency

C   Harmonics occur at exact (integer) multiples of the fundamental frequency. In other words, the harmonics are exactly two, or three, four, or more times the desired frequency.

T2A03 What does 60 hertz (Hz) mean?
- A. 6000 cycles per second
- B. 60 cycles per second
- C. 6000 meters per second
- D. 60 meters per second

B   We measure frequency in hertz (abbreviated Hz). Frequency is a measure of the number of times in one second the alternating current flows back and forth. One cycle per second is 1 Hz. 60 cycles per second is 60 Hz.

T2A04 What is the name for the distance an AC signal travels during one complete cycle?
- A. Wave speed
- B. Waveform
- C. Wavelength
- D. Wave spread

C   The distance an AC signal travels during one complete cycle is called the signal's wavelength.

T2A05 What is the fourth harmonic of a 50.25 MHz signal?
- A. 201.00 MHz
- B. 150.75 MHz
- C. 251.50 MHz
- D. 12.56 MHz

A   The fourth harmonic of a signal is exactly four times the fundamental (desired) frequency. That means that the fourth harmonic of 50.25 MHz is: 50.25 × 4 = 201.0 MHz

T2A06 What is a radio frequency wave?

A. Wave disturbances that take place at less than 10 times per second

B. Electromagnetic oscillations or cycles that repeat between 20 and 20,000 times per second

C. Electromagnetic oscillations or cycles that repeat more than 20,000 times per second

D. None of these answers are correct

C   Radio frequency waves are electromagnetic waves or cycles that repeat more than 20,000 times per second. Radio frequencies are higher than audio frequencies.

T2A07 What is an audio-frequency signal?

A. Wave disturbances that cannot be heard by the human ear

B. Electromagnetic oscillations or cycles that repeat between 20 and 20,000 times per second

C. Electromagnetic oscillations or cycles that repeat more than 20,000 times per second

D. Electric energy that is generated at the front end of an AM or FM radio receiver

B   An audio-frequency signal is comprised of electromagnetic waves or cycles that repeat between 20 and 20,000 times per second. Audio frequencies are lower than radio frequencies.

T2A08 In what radio-frequency range do amateur 2-meter communications take place?

A. UHF, Ultra High Frequency range

B. MF, Medium Frequency range

C. HF, High Frequency range

D. VHF, Very High Frequency range

D   The 2-meter amateur band is from 144 to 148 MHz. VHF is the range from 30 to 300 MHz, which is where you'll find the 2-meter band. **Table 2-1** shows the limits for common frequency ranges.

Table 2-1
Frequency Ranges

Designation	Abbreviation	Range
Medium	MF	0.3 – 3 MHz
High	HF	3 – 30 MHz
Very High	VHF	30 – 300 MHz
Ultra High	UHF	300 – 3000 MHz

T2A09 Which of the following choices is often used to identify a particular radio wave?

    A. The frequency or the wavelength of the wave
    B. The length of the magnetic curve of wave
    C. The time it takes for the wave to travel a certain distance
    D. The free-space impedance of the wave

A   Hams frequently identify a particular radio wave by its frequency. They might say something like, "that CW on 3.5185 MHz is W1AW code practice." They also use wavelength. An example would be, "I can't hear 80-meter code practice from W1AW at my house, but it comes in very strong on 40 meters."

T2A10 How is a radio frequency wave identified?

    A. By its wavelength, the length of a single radio cycle from peak to peak
    B. By its corresponding frequency
    C. By the appropriate radio band in which it is transmitted or received
    D. All of these choices are correct

D   All of these choices are correct. If you need to, go back and read the explanation to the previous question.

T2A11 How fast does a radio wave travel through space (in a vacuum)?

    A. At the speed of light
    B. At the speed of sound
    C. Its speed is inversely proportional to its wavelength
    D. Its speed increases as the frequency increases

A   A radio wave travels through the vacuum of space at the speed of light. The speed of light is 300,000,000 meters per second (3.00×10^8 m/s).

T2A12 What is the standard unit of frequency measurement?

- A. A megacycle
- B. A hertz
- C. One thousand cycles per second
- D. EMF, electromagnetic force

B   The standard unit of frequency measurement is the hertz. A signal with a frequency of one hertz completes one cycle in one second.

T2A13 What is the basic principle of radio communications?

- A. A radio wave is combined with an information signal and is transmitted; a receiver separates the two
- B. A transmitter separates information to be received from a radio wave
- C. A DC generator combines some type of information into a carrier wave so that it may travel through space
- D. The peak-to-peak voltage of a transmitter is varied by the sidetone and modulated by the receiver

A   The basic principle of radio communications involves transmitting and receiving information, such as your voice. You speak into a microphone and that signal is combined with a carrier wave. The combined signals are transmitted to a receiver where your voice is separated from the carrier. See the drawing on the next page.

T2A14 How is the wavelength of a radio wave related to its frequency?

- A. Wavelength gets longer as frequency increases
- B. Wavelength gets shorter as frequency increases
- C. There is no relationship between wavelength and frequency
- D. The frequency depends on the velocity of the radio wave, but the wavelength depends on the bandwidth of the signal

B   The wavelength of a radio wave is inversely related to its frequency. That means that as the frequency of a radio wave goes up, the wavelength (distance traveled during one cycle) goes down. If the frequency is doubled, the wavelength is cut in half.

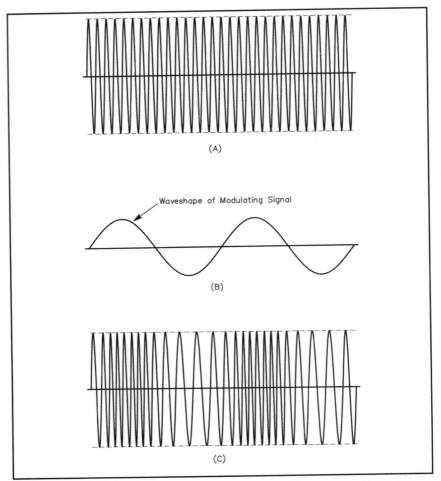

This drawing shows a graphical representation of frequency modulation. In the unmodulated carrier at A, each RF cycle takes the same amount of time to complete. When the modulating signal at B is applied, the carrier frequency is increased or decreased according to the amplitude and polarity of the modulating signal, as shown at C.

T2A15 What term means the number of times per second that an alternating current flows back and forth?

- A. Pulse rate
- B. Speed
- C. Wavelength
- D. Frequency

D   The frequency of an alternating current is the number of times per second that an alternating current flows back and forth.

T2A16 What is the basic unit of frequency?

A. The hertz
B. The watt
C. The ampere
D. The ohm

A The basic unit of frequency is the hertz. A signal with a frequency of one hertz completes one cycle in one second.

T2B **Frequency privileges granted to Technician class operators; Amateur service bands; Emission types and designators; Modulation principles; AM/FM/ Single sideband/upper-lower, international Morse code (CW), RTTY, packet radio and data emission types; Full quieting.**

T2B01 What are the frequency limits of the 80-meter band in ITU Region 2 for Technician class licensees who have passed a Morse code exam?

A. 3500 - 4000 kHz
B. 3675 - 3725 kHz
C. 7100 - 7150 kHz
D. 7000 - 7300 kHz

B Technician class licensees who have passed a Morse code exam may operate on the 80-meter band using CW only between 3675 and 3725 kHz. **Table 2-2** is a handy reference to help you with this and similar questions in this section. [97.301 (e)]

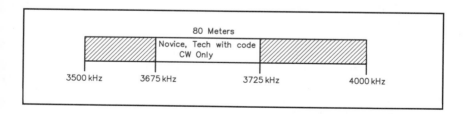

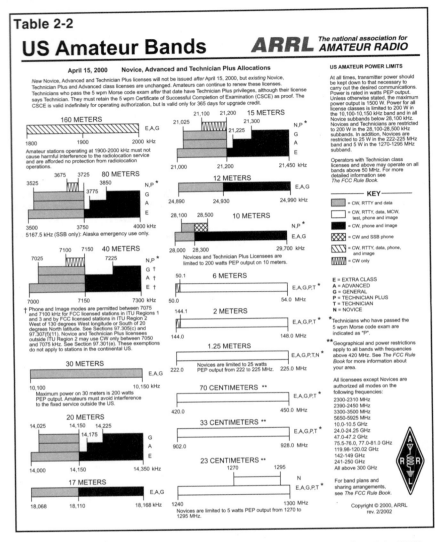

Table 2-2

US Amateur Bands

ARRL *The national association for* **AMATEUR RADIO**

April 15, 2000 Novice, Advanced and Technician Plus Allocations

New Novice, Advanced and Technician Plus licenses will not be issued *after* April 15, 2000, but *existing* Novice, Technician Plus and Advanced class licenses are unchanged. Amateurs can continue to renew these licenses. Technicians who pass the 5 wpm Morse code exam *after* that date have Technician Plus privileges, although their license says Technician. They must retain the 5 wpm Certificate of Successful Completion of Examination (CSCE) as proof. The CSCE is valid indefinitely for operating authorization, but is valid only for 365 days for upgrade credit.

160 METERS

1800 1900 2000 kHz — E,A,G

Amateur stations operating at 1900-2000 kHz must not cause harmful interference to the radiolocation service and are afforded no protection from radiolocation operations.

80 METERS
3525 3675 3725 3850 3775 — N,P *, G, A, E
3500 3750 4000 kHz
5167.5 kHz (SSB only): Alaska emergency use only.

40 METERS
7025 7100 7150 7225 — N,P *, G †, A †, E †
7000 7150 7300 kHz

† Phone and Image modes are permitted between 7075 and 7100 kHz for FCC licensed stations in ITU Regions 1 and 3 and by FCC licensed stations in ITU Region 2 West of 130 degrees West longitude or South of 20 degrees North latitude. See Sections 97.305(c) and 97.307(f)(11). Novice and Technician Plus licensees outside ITU Region 2 may use CW only between 7050 and 7075 kHz. See Section 97.301(e). These exemptions do not apply to stations in the continental US.

30 METERS
10,100 10,150 kHz — E,A,G
Maximum power on 30 meters is 200 watts PEP output. Amateurs must avoid interference to the fixed service outside the US.

20 METERS
14,025 14,150 14,225 14,175 — G, A, E
14,000 14,150 14,350 kHz

17 METERS
18,068 18,110 18,168 kHz — E,A,G

15 METERS
21,025 21,100 21,200 21,300 21,225 — N,P *, G, A, E
21,000 21,200 21,450 kHz

12 METERS
24,890 24,930 24,990 kHz — E,A,G

10 METERS
28,100 28,500 28,300 — N,P *, E,A,G
28,000 28,300 29,700 kHz
Novices and Technician Plus Licensees are limited to 200 watts PEP output on 10 meters.

6 METERS
50.1 50.0 54.0 MHz — E,A,G,P,T *

2 METERS
144.1 144.0 148.0 MHz — E,A,G,P,T *

1.25 METERS
222.0 225.0 MHz — E,A,G,P,T,N *
Novices are limited to 25 watts PEP output from 222 to 225 MHz.

70 CENTIMETERS *
420.0 450.0 MHz — E,A,G,P,T *

33 CENTIMETERS *
902.0 928.0 MHz — E,A,G,P,T *

23 CENTIMETERS *
1270 1295 — N
1240 1300 MHz — E,A,G,P,T *
Novices are limited to 5 watts PEP output from 1270 to 1295 MHz.

US AMATEUR POWER LIMITS

At all times, transmitter power should be kept down to that necessary to carry out the desired communications. Power is rated in watts PEP output. Unless otherwise stated, the maximum power output is 1500 W. Power for all license classes is limited to 200 W in the 10,100-10,150 kHz band and in all Novice subbands below 28,100 kHz. Novices and Technicians are restricted to 200 W in the 28,100-28,500 kHz subbands. In addition, Novices are restricted to 25 W in the 222-225 MHz band and 5 W in the 1270-1295 MHz subband.

Operators with Technician class licenses and above may operate on all bands above 50 MHz. For more detailed information see *The FCC Rule Book.*

— KEY —

▨ = CW, RTTY and data

☐ = CW, RTTY, data, MCW, test, phone and image

■ = CW, phone and image

▨▨ = CW and SSB phone

⧅⧅ = CW, RTTY, data, phone, and image

▨▨▨ = CW only

E = EXTRA CLASS
A = ADVANCED
G = GENERAL
P = TECHNICIAN PLUS
T = TECHNICIAN
N = NOVICE

* Technicians who have passed the 5 wpm Morse code exam are indicated as "P".

** Geographical and power restrictions apply to all bands with frequencies above 420 MHz. See *The FCC Rule Book* for more information about your area.

All licensees except Novices are authorized all modes on the following frequencies:
2300-2310 MHz
2390-2450 MHz
3300-3500 MHz
5650-5925 MHz
10.0-10.5 GHz
24.0-24.25 GHz
47.0-47.2 GHz
75.5-76.0, 77.0-81.0 GHz
119.98-120.02 GHz
142-149 GHz
241-250 GHz
All above 300 GHz

For band plans and sharing arrangements, see *The FCC Rule Book.*

Copyright © 2000, ARRL
rev. 2/2002

T2B02 What are the frequency limits of the 10-meter band in ITU Region 2 for Technician class licensees who have passed a Morse code exam?

    A. 28.000 - 28.500 MHz
    B. 28.100 - 29.500 MHz
    C. 28.100 - 28.500 MHz
    D. 29.100 - 29.500 MHz

  C   Technician class licensees who have passed a Morse code exam may operate on the 10-meter band between 28.100 and 28.500 MHz. They may operate CW, RTTY and data modes between 28.100 and 28.300. They also have CW and SSB phone privileges between 28.300 and 28.500. [97.301 (e)]

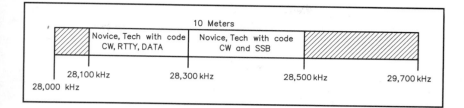

T2B03 What name does the FCC use for telemetry, telecommand or computer communications emissions?

A. CW
B. Image
C. Data
D. RTTY

C   The FCC uses the term "data" to refer to telemetry, telecommand or computer communications emissions. Further technical details can be found in the rules. [97.3 (c) (2)]

T2B04 What does "connected" mean in a packet-radio link?

A. A telephone link is working between two stations
B. A message has reached an amateur station for local delivery
C. A transmitting station is sending data to only one receiving station; it replies that the data is being received correctly
D. A transmitting and receiving station are using a digipeater, so no other contacts can take place until they are finished

C   When operating packet radio, "connected" means that a transmitting station is sending data to only one receiving station. That receiving station replies that the data is being received correctly. In other words, a packet-radio link is point-to-point communications. It is not a broadcast to multiple station.

T2B05 What emission types are Technician control operators who have passed a Morse code exam allowed to use from 7100 to 7150 kHz in ITU Region 2?

    A. CW and data
    B. Phone
    C. Data only
    D. CW only

    D   Technician class licensees in ITU Region 2 who have passed a Morse code exam may operate CW only on the 40-meter band between 7100 to 7150 kHz. [97.305, 97.307 (f) (9)]

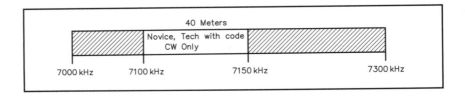

T2B06 What emission types are Technician control operators who have passed a Morse code exam allowed to use on frequencies from 28.3 to 28.5 MHz?

    A. All authorized amateur emission privileges
    B. CW and data
    C. CW and single-sideband phone
    D. Data and phone

    C   Technician control operators who have passed a Morse code exam are allowed to use CW and single-sideband phone on frequencies from 28.3 to 28.5 MHz. [97.305, 97.307 (f) (10)]

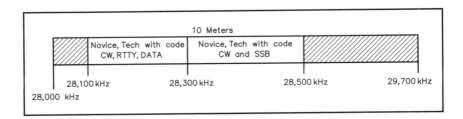

T2B07 What emission types are Technician control operators allowed to use on the amateur 1.25-meter band in ITU Region 2?

A. Only CW and phone
B. Only CW and data
C. Only data and phone
D. All amateur emission privileges authorized for use on the band

D   Technician control operators in ITU Region 2 are allowed all emission privileges authorized for use on the amateur 1.25-meter band. [97.305]

1.25 METERS

222.0   Novices are limited to 25 watts PEP output on 1.25 meters. 225.0 MHz

▓ =CW, RTTY, data, MCW, test, phone and image

T2B08 What term describes the process of combining an information signal with a radio signal?

A. Superposition
B. Modulation
C. Demodulation
D. Phase-inversion

B   To use a radio signal to communicate information, you'll have to add the information to the signal somehow. This process of combining an information signal with a radio signal is called modulation. The simplest form of modulation is to turn the radio signal on and off. CW, also called international Morse code, is transmitted by on/off keying of a radio-frequency signal.

T2B09 What is the name of the voice emission most used on VHF/UHF repeaters?

A. Single-sideband phone
B. Pulse-modulated phone
C. Slow-scan phone
D. Frequency-modulated phone

D   Nearly all VHF and UHF voice repeaters use FM phone. A practical FM transmitter varies the carrier frequency or phase to produce the modulated signal.

T2B10 What does the term "phone transmissions" usually mean?

A. The use of telephones to set up an amateur contact
B. A phone patch between amateur radio and the telephone system
C. AM, FM or SSB voice transmissions by radiotelephony
D. Placing the telephone handset near a transceiver's microphone and speaker to relay a telephone call

C   Any voice mode used for communication is known as a phone emission under FCC Rules. AM, SSB and FM voice are all phone emission types.

T2B11 Which sideband is commonly used for 10-meter phone operation?

A. Upper sideband
B. Lower sideband
C. Amplitude-compandored sideband
D. Double sideband

A   You can use either the lower sideband or the upper sideband to transmit an SSB signal. Amateurs normally use the upper sideband for 10-meter phone operation.

T2B12 What is the most transmitter power a Technician control operator with telegraphy credit may use on the 10-meter band?

A. 5 watts PEP output
B. 25 watts PEP output
C. 200 watts PEP output
D. 1500 watts PEP output

C   Technician operators with Morse code credit are limited to 200 watts PEP output on the HF bands. [97.313 (c) (2)]

T2B13 What name does the FCC use for voice emissions?

A. RTTY
B. Data
C. CW
D. Phone

D   Any voice mode used for communication is known as a phone emission under FCC Rules. AM, SSB and FM voice are all phone emission types. [97.3 (c) (5)]

T2B14 What emission privilege is permitted a Technician class operator in the 219 MHz - 220 MHz frequency range?

A. Slow-scan television
B. Point-to-point digital message forwarding
C. FM voice
D. Fast-scan television

B   A Technician class operator is allowed to use point-to-point digital message forwarding between 219 MHz and 220 MHz. Amateurs are not allowed any other emission types in this part of the band. See the rules for other limitations before you begin operations. [97.305 (c)]

T2B15 Which sideband is normally used for VHF/UHF SSB communications?

A. Upper sideband
B. Lower sideband
C. Double sideband
D. Double sideband, suppressed carrier

A   Normally, hams use upper sideband for VHF/UHF SSB communications.

T2B16 Which of the following descriptions is used to describe a good signal through a repeater?

A. Full quieting
B. Over deviation
C. Breaking up
D. Readability zero

A   Of the choices given, only "full quieting" describes a good signal. When an FM signal is at or above a certain level at the receiver, no noise will be heard on that signal. A signal that is strong enough to cause the noise to disappear is said to be "full quieting."

T2B17 This question has been withdrawn.

T2B18 What emissions do a transmitter using a reactance modulator produce?

    A. CW
    B. Test
    C. Single-sideband, suppressed-carrier phone
    D. Phase-modulated phone

D   A reactance modulator modulates (shifts) the phase of an RF oscillator according to an applied signal. This results in a phase-modulated output.

T2B19 What other emission does phase modulation most resemble?

    A. Amplitude modulation
    B. Pulse modulation
    C. Frequency modulation
    D. Single-sideband modulation

C   Phase modulation is closely related to frequency modulation. You can not modulate a signals phase without shifting its frequency and vice versa.

Radio Phenomena

There will be 2 questions on your exam taken from the Radio Phenomena subelement printed in this chapter. These questions are divided into 2 groups, labeled T3A and T3B

T3A   How a radio signal travels; Atmosphere/troposphere/ ionosphere and ionized layers; Skip distance; Ground (surface)/sky (space) waves; Single/multihop; Path; Ionospheric absorption; Refraction.

T3A01 What is the name of the area of the atmosphere that makes long-distance radio communications possible by bending radio waves?

A. Troposphere
B. Stratosphere
C. Magnetosphere
D. Ionosphere

D   Ultraviolet radiation from the sun produces charged molecules (ions) in the portion of the outer atmosphere called the **ionosphere**. These ionized particles allow high-frequency (HF) radio signals to travel large distances by bending them back to earth.

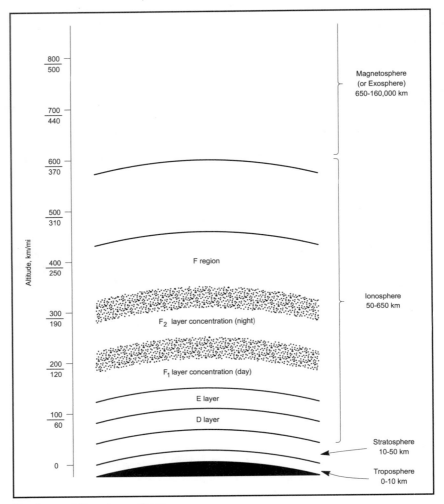

Regions of the atmosphere.

T3A02 Which ionospheric region is closest to the Earth?

    A. The A region
    B. The D region
    C. The E region
    D. The F region

B   The lowest region of the ionosphere affecting propagation is the **D region**. This region is in a relatively dense part of the atmosphere about 35 to 60 miles above the Earth.

T3A03 Which region of the ionosphere is mainly responsible for absorbing MF/HF radio signals during the daytime?

A. The F2 region
B. The F1 region
C. The E region
D. The D region

D   The **D region** of the ionosphere exists only during the daytime, and it absorbs long-wavelength (low-frequency) radio signals that try to pass through it. Signals in the amateur 160, 80 and 40-meter (MF/HF) bands are most affected.

T3A04 Which region of the ionosphere is mainly responsible for long-distance sky-wave radio communications?

A. D region
B. E region
C. F1 region
D. F2 region

D   The **F region** is the portion of the ionosphere most responsible for long-distance amateur communication on the HF bands. A one-hop radio transmission travels a maximum of about 2500 miles using the F2 region, the upper part of the F region.

T3A05 When a signal travels along the surface of the Earth, what is this called?

A. Sky-wave propagation
B. Knife-edge diffraction
C. E-region propagation
D. Ground-wave propagation

D   Radio signals follow the surface of the Earth, even going over low hills in some cases, producing **ground-wave propagation**. Low-frequency, long-wavelength signals travel farthest along the ground; as far as 100 miles for stations near the low-frequency end of the standard AM broadcast band (the 540-kHz end). Higher-frequency, shorter-wavelength signals cannot be heard as far away. On the 80-meter Novice band, signals may travel up to about 70 miles by ground wave. At 28 MHz, in the amateur 10-meter band, you won't hear ground-wave signals more than 10 to 15 miles away.

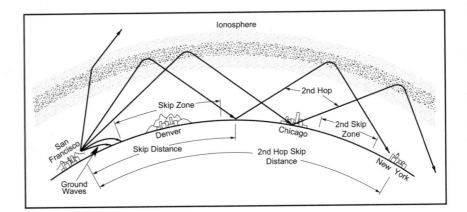

T3A06 What type of solar radiation is most responsible for ionization in the outer atmosphere?

    A. Thermal
    B. Non-ionized particle
    C. Ultraviolet
    D. Microwave

C   **Ultraviolet radiation** from the sun is most responsible for the ionization in the outer atmosphere. (Ultraviolet radiation has a higher frequency — shorter wavelength — than visible light.)

T3A07 What is the usual cause of sky-wave propagation?

    A. Signals are reflected by a mountain
    B. Signals are reflected by the Moon
    C. Signals are bent back to Earth by the ionosphere
    D. Signals are retransmitted by a repeater

C   When ionized by solar radiation, the ionosphere can refract (bend) radio waves. If the wave is bent enough, it returns to Earth as **sky-wave propagation**. If the wave is not bent enough, it is lost as it travels off into space.

T3A08 What type of propagation has radio signals bounce several times between Earth and the ionosphere as they travel around the Earth?

- A. Multiple bounce
- B. Multi-hop
- C. Skip
- D. Pedersen propagation

B   Ham radio contacts of up to 2500 miles are possible with one skip off the ionosphere. Worldwide communications using several skips (or **multi-hops**) can take place if conditions are right, especially when radio waves are launched at low takeoff angles into the ionosphere. This is the way long-distance radio signals travel.

T3A09 What effect does the D region of the ionosphere have on lower-frequency HF signals in the daytime?

- A. It absorbs the signals
- B. It bends the radio waves out into space
- C. It refracts the radio waves back to earth
- D. It has little or no effect on 80-meter radio waves

A   The **D region** of the ionosphere, which exists only during the daytime, absorbs long-wavelength (low-frequency) radio signals that try to pass through it. Signals in the amateur 160, 80 and 40-meter (MF/HF) bands are most affected.

T3A10 How does the signal loss for a given path through the troposphere vary with frequency?

- A. There is no relationship
- B. The path loss decreases as the frequency increases
- C. The path loss increases as the frequency increases
- D. There is no path loss at all

C   As radio waves travel through the troposphere, there will always be some signal loss. For any particular path through the troposphere, the overall **path loss** increases as the frequency increases.

T3A11 When a signal is returned to Earth by the ionosphere, what is this called?

A. Sky-wave propagation
B. Earth-Moon-Earth propagation
C. Ground-wave propagation
D. Tropospheric propagation

A   When ionized by solar radiation, the ionosphere can refract (bend) radio waves. If the wave is bent enough, it returns to Earth, giving us **sky-wave propagation**. If the wave is not bent enough, it is lost as it travels off into space.

T3A12 How does the range of sky-wave propagation compare to ground-wave propagation?

A. It is much shorter
B. It is much longer
C. It is about the same
D. It depends on the weather

B   In **ground-wave propagation**, radio waves travel along the Earth's surface. During the day you might have an 80-meter contact with a station as much as 100 miles away by means of ground waves. But during the day you might be able to speak with someone thousands of miles away by means of **sky-wave propagation**.

T3B   HF vs. VHF vs. UHF characteristics; Types of VHF-UHF propagation; Daylight and seasonal variations; Tropospheric ducting; Line of sight; Maximum usable frequency (MUF); Sunspots and sunspot Cycle, Characteristics of different bands.

T3B01 When a signal travels in a straight line from one antenna to another, what is this called?

A. Line-of-sight propagation
B. Straight line propagation
C. Knife-edge diffraction
D. Tunnel ducting

A   Radio waves travel in a straight line from a transmitting antenna, but they can bend as they travel along the Earth's surface or into the ionosphere. Radio waves also reflect, or bounce off, objects in their paths. When radio waves travel directly from the transmitting antenna to the receiving antenna in a straight line, with no bending or reflection, we call it **line-of-sight propagation**. If you and a friend are using 70-cm hand-held radios to communicate directly with each other over a distance of a few miles, you are using line-of-sight propagation. This is the most common form of propagation when you are using VHF or UHF radios.

T3B02 What can happen to VHF or UHF signals going towards a metal-framed building?

    A. They will go around the building
    B. They can be bent by the ionosphere
    C. They can be reflected by the building
    D. They can be polarized by the building's mass

    C   VHF and UHF signals are easily **reflected by buildings,** mountains and other objects in their paths. If there is something in the way that blocks a direct signal from your station to a friend's station, you might be able to bounce your signals off a building or mountain. In that case, you may have to point your beam (directional) antenna away from your friend's location and toward the building or mountain to produce the best signals.

T3B03 Ducting occurs in which region of the atmosphere?

    A. F2
    B. Ecosphere
    C. Troposphere
    D. Stratosphere

    C   The troposphere is a layer of the atmosphere that is below the ionosphere. When a warm air mass covers a cold air mass in the troposphere, a **duct** may form that traps radio waves in the cold air.

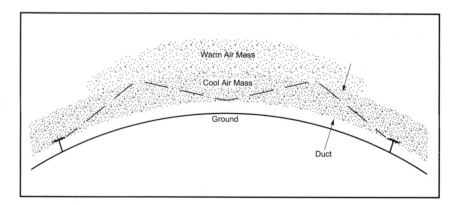

T3B04 What causes VHF radio waves to be propagated several hundred miles over oceans?

    A. A polar air mass
    B. A widespread temperature inversion
    C. An overcast of cirriform clouds
    D. A high-pressure zone

B   **A widespread temperature inversion** can propagate VHF radio waves several hundred miles. Such temperature inversions usually occur over oceans, although they can also occur over land. Also see the explanation for T3B03.

T3B05 In which of the following frequency ranges does sky-wave propagation least often occur?

    A. LF
    B. UHF
    C. HF
    D. VHF

B   **UHF** radio waves do not travel by means of **sky-wave propagation** through the ionosphere. UHF waves only propagate through the troposphere, traveling perhaps 15 percent farther away than the true horizon because of slight bending in the troposphere.

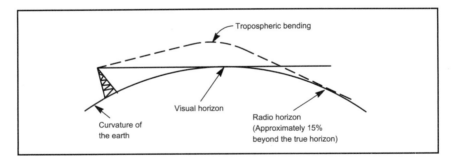

T3B06 Why should local amateur communications use VHF and UHF frequencies instead of HF frequencies?

    A. To minimize interference on HF bands capable of long-distance communication
    B. Because greater output power is permitted on VHF and UHF
    C. Because HF transmissions are not propagated locally
    D. Because signals are louder on VHF and UHF frequencies

A   Because HF radio waves can so easily travel long distances, even with low transmitter power, short-range communication is best carried out on the VHF and UHF bands. This prevents or minimizes interference to other hams using the HF bands. In particular, UHF radio waves do not travel by means of sky wave propagation through the ionosphere.

T3B07 How does the number of sunspots relate to the amount of ionization in the ionosphere?

    A. The more sunspots there are, the greater the ionization
    B. The more sunspots there are, the less the ionization
    C. Unless there are sunspots, the ionization is zero
    D. Sunspots do not affect the ionosphere

    A   Sunspots are grayish-black blotches on the sun's surface. **More sunspots usually means more ionization** of the ionosphere. As a result, the higher-frequency bands tend to open for longer-distance communications when there are more spots.

T3B08 How long is an average sunspot cycle?

    A. 2 years
    B. 5 years
    C. 11 years
    D. 17 years

    C   The number and size of sunspots vary over approximately an **11-year cycle**. During a peak in sunspot activity you will often be able to communicate all over the world using a few watts of transmitter power on the 10-meter band. During the sunspot minimum you will have to move to lower-frequency bands like 40 and 80 meters for reliable worldwide communications.

T3B09 Which of the following frequency bands is most likely to experience summertime sporadic-E propagation?

    A. 23 centimeters
    B. 6 meters
    C. 70 centimeters
    D. 1.25 meters

    B   **Sporadic E**, or "E skip," is a type of sky-wave propagation that allows long-distance communications on the VHF bands (6 meters, 2 meters and 222 MHz) through the E region of the ionosphere. Although, as its name implies, sporadic E occurs only sporadically, during certain times of the year, in the summer, it is the most common type of VHF ionospheric propagation.

T3B10 Which of the following emission modes are considered to be weak-signal modes and have the greatest potential for DX contacts?

   A. Single sideband and CW
   B. Packet radio and RTTY
   C. Frequency modulation
   D. Amateur television

A   Amateurs commonly use SSB or CW as **weak-signal modes** for VHF/UHF **DX** work — as opposed to "strong-signal modes," such as local FM or packet radio.

T3B11 What is the condition of the ionosphere above a particular area of the Earth just before local sunrise?

   A. Atmospheric attenuation is at a maximum
   B. The D region is above the E region
   C. The E region is above the F region
   D. Ionization is at a minimum

D   Both the E and F regions reach maximum levels of ionization around midday. By early evening the ionization level in the E region is very low and its ionization level reaches a **minimum** just before **sunrise**, local time. In the F region, ionization tapers off very gradually toward sunset after noon, since at this altitude the ions and electrons recombine very slowly because there aren't many of them around. The F region thus remains ionized during the night, also reaching a **minimum** just before **sunrise**.

T3B12 What happens to signals that take off vertically from the antenna and are higher in frequency than the critical frequency?

   A. They pass through the ionosphere
   B. They are absorbed by the ionosphere
   C. Their frequency is changed by the ionosphere to be below the maximum usable frequency
   D. They are reflected back to their source

A   The highest frequency where a radio wave transmitted straight upwards into the ionosphere will be reflected back down to the Earth is called the **critical frequency**. If you raise the frequency above the critical frequency a wave launched straight up will be lost into space.

T3B13 In relation to sky-wave propagation, what does the term "maximum usable frequency" (MUF) mean?

- A. The highest frequency signal that will reach its intended destination
- B. The lowest frequency signal that will reach its intended destination
- C. The highest frequency signal that is most absorbed by the ionosphere
- D. The lowest frequency signal that is most absorbed by the ionosphere

A   The maximum usable frequency (MUF) is the **highest-frequency signal** that is bent back to Earth to provide communications **between two specific locations**. The MUF for different propagation paths may be very different.

Station Licensee Duties

There will be 3 questions on your exam taken from the Station Licensee Duties subelement printed in this chapter. These questions are divided into 3 groups, labeled T4A, T4B and T4C.

After some of the explanations in this chapter you will see a reference to Part 97 of the Federal Communications Commission Rules, set inside square brackets, like [97.3a5]. This tells you where to look for the exact wording of the Rules as they relate to that question. For a complete copy of Part 97, along with simple explanations of the Rules governing Amateur Radio, obtain a copy of *The ARRL's FCC Rule Book*.

T4A **Correct name and mailing address on station license grant; Places from where station is authorized to transmit; Selecting station location; Antenna structure location; Stations installed aboard ship or aircraft.**

T4A01 When may you operate your amateur station aboard a cruise ship?
A. At any time
B. Only while the ship is not under power
C. Only with the approval of the master of the ship and not using the ship's radio equipment
D. Only when you have written permission from the cruise line and only using the ship's radio equipment

C   Someday, you may want to operate your equipment aboard a ship. To do that you must have the permission of the ship's master and you must not use the ship's radio equipment. In addition, your station must not constitute a hazard to the safety of life or property. [97.11 (a)]

T4A02 When may you operate your amateur station somewhere in the US besides the address listed on your license?

A. Only during times of emergency
B. Only after giving proper notice to the FCC
C. During an emergency or an FCC-approved emergency practice
D. Whenever you want to

D You may operate your amateur station at a location in the US other than the address listed on your license whenever you want to. The rules do not limit the "when."

T4A03 What penalty may the FCC impose if you fail to provide your correct mailing address?

A. There is no penalty if you do not provide the correct address
B. You are subject to an administrative fine
C. Your amateur license could be revoked
D. You may only operate from your address of record

C The FCC rules specifically state that your amateur license could be revoked if mail from FCC is returned to them as undeliverable because you failed to supply the correct address. Remember to change your address whenever you move your residence. [97.23]

T4A04 Under what conditions may you transmit from a location different from the address printed on your amateur license?

A. If the location is under the control of the FCC, whenever the FCC Rules allow
B. If the location is outside the United States, only for a time period of less than 90 days
C. Only when you have written permission from the FCC Engineer in Charge
D. Never; you may only operate at the location printed on your license

A You may transmit from a location different than the address printed on your amateur license under certain conditions. Those conditions are that the location is an area regulated by the FCC, and that the operation is in accordance with the FCC Rules.

T4A05 Why must an amateur operator have a current US postal mailing address?

    A. So the FCC has a record of the location of each amateur station
    B. To follow the FCC rules and so the licensee can receive mail from the FCC
    C. Because all US amateurs must be US residents
    D. So the FCC can publish a call-sign directory

B   As an amateur operator, you must have a current US postal mailing address. The FCC rules require it. Further, the rules specifically state that your amateur license could be revoked if mail from FCC is returned to them as undeliverable because you failed to supply the correct address. [97.23]

T4A06 What is one way to notify the FCC if your mailing address changes?

    A. Fill out an FCC Form 605 using your new address, attach a copy of your license, and mail it to your local FCC Field Office
    B. Fill out an FCC Form 605 using your new address, attach a copy of your license, and mail it to the FCC office in Gettysburg, PA
    C. Call your local FCC Field Office and give them your new address over the phone
    D. Call the FCC office in Gettysburg, PA, and give them your new address over the phone

B   You can notify the FCC of a mailing address change on an FCC Form 605. Use your new address, attach a copy of your license, and mail it to the FCC office in Gettysburg, PA. [97.21 (a) (1)]

T4A07 What do FCC rules require you to do if you plan to erect an antenna whose height exceeds 200 feet?

    A. Notify the Federal Aviation Administration and register with the FCC
    B. FCC rules prohibit antenna structures above 200 feet
    C. Alternating sections of the supporting structure must be painted international airline orange and white
    D. The antenna structure must be approved by the FCC and DOD

A   A restriction designed to protect others is the antenna-height limitation. You may not build an antenna structure (this includes the radiating elements, tower supports or any other attachments) that is over 200 feet high, unless you first notify the Federal Aviation Administration (FAA) and register with the FCC. [97.15 (a)]

T4A08 Which of the following is NOT an important consideration when selecting a location for a transmitting antenna?

A. Nearby structures
B. Height above average terrain
C. Distance from the transmitter location
D. Polarization of the feed line

D   The location of nearby structures, the antenna height above average terrain, and the distance from the transmitter location are all important considerations when selecting a location for a transmitting antenna. "Polarization of the feed line" is a nonsense phrase. [97.13 (c)] [OET Bulletin 65 Supplement B] ["*RF Exposure and You*", W1RFI]

T4A09 What is the height restriction the FCC places on Amateur Radio Service antenna structures without registration with the FCC and FAA?

A. There is no restriction by the FCC
B. 200 feet
C. 300 feet
D. As permitted by PRB-1

B   If you want to erect an antenna whose height exceeds 200 feet, FCC rules require you to notify the Federal Aviation Administration and register with the FCC. [97.15 (b)]

T4A10 When may you operate your amateur station aboard an aircraft?

A. At any time
B. Only while the aircraft is on the ground
C. Only with the approval of the pilot in command and not using the aircraft's radio equipment
D. Only when you have written permission from the airline and only using the aircraft's radio equipment

C   Someday, you may want to operate your equipment aboard an aircraft. To do that you must have the permission of the pilot in command of the aircraft, and you must not use the aircraft's radio equipment. You are not likely to receive permission to operate on board a commercial airplane. [97.11 (a)]

T4B **Designation of control operator; FCC presumption of control operator; Physical control of station apparatus; Control point; Immediate station control; Protecting against unauthorized transmissions; Station records; FCC Inspection; Restricted operation.**

T4B01 What is the definition of a control operator of an amateur station?

- A. Anyone who operates the controls of the station
- B. Anyone who is responsible for the station's equipment
- C. Any licensed amateur operator who is responsible for the station's transmissions
- D. The amateur operator with the highest class of license who is near the controls of the station

C   A control operator is an "amateur operator designated by the licensee of a station to be responsible for the transmissions from that station to assure compliance with the FCC Rules." In effect, the control operator operates the Amateur Radio station. Only a licensed ham may be the control operator of an amateur station. If another licensed radio amateur operates your station with your permission, he or she assumes the role of control operator. [97.3 (a) (12)]

T4B02 What is the FCC's name for the person responsible for the transmissions from an amateur station?

- A. Auxiliary operator
- B. Operations coordinator
- C. Third-party operator
- D. Control operator

D   The control operator is the licensed ham responsible for the transmissions from an amateur station. [97.3 (a) (12)]

T4B03 When must an amateur station have a control operator?

- A. Only when training another amateur
- B. Whenever the station receiver is operated
- C. Whenever the station is transmitting
- D. A control operator is not needed

C   An amateur station must have a control operator whenever the station is transmitting. In other words at all times that the station is transmitting. [97.7]

T4B04 What is the term for the location at which the control operator function is performed?

    A. The operating desk
    B. The control point
    C. The station location
    D. The manual control location

B   The control operator function is performed at the control point. [97.3 (a) (13)]

T4B05 What is the control point of an amateur station?

    A. The on/off switch of the transmitter
    B. The input/output port of a packet controller
    C. The variable frequency oscillator of a transmitter
    D. The location at which the control operator function is performed

D   The control point of an amateur station is the location at which the control operator function is performed. [97.3 (a) (13)]

T4B06 When you operate your transmitting equipment alone, what is your official designation?

    A. Engineer in Charge
    B. Commercial radio operator
    C. Third party
    D. Control operator

D   When you are the lone operator of your transmitting equipment, you are the control operator. That is your official designation. [97.3 (a) (12)]

T4B07 When does the FCC assume that you authorize transmissions with your call sign as the control operator?

A. At all times
B. Only in the evening hours
C. Only when operating third party traffic
D. Only when operating as a reciprocal operating station

A   Any amateur operator may designate another licensed operator as the control operator, to share the responsibility of station operation. However, the FCC holds both the control operator and the station licensee responsible for proper operation of the station. The FCC rules specifically state that they (FCC) assume that you are the control operator whenever your call sign is used on-the-air. [97.103 (b)]

T4B08 What is the name for the operating position where the control operator has full control over the transmitter?

A. Field point
B. Auxiliary point
C. Control point
D. Access point

C   A control operator must be present at the station control point whenever a transmitter is operating. The FCC defines control point as "the location at which the control operator function is performed." [97.3 (a) (13)]

T4B09 When is the FCC allowed to conduct an inspection of your amateur station?

A. Only on weekends
B. At any time
C. Never, the FCC does not inspect stations
D. Only during daylight hours

B   The FCC can conduct an inspection of your amateur station at any time. You must make both the station and any station records available for inspection upon request by an FCC representative. [97.103 (c)]

T4B10 How many transmitters may an amateur licensee control at the same time?

A. Only one
B. No more than two
C. Any number
D. Any number, as long as they are transmitting in different bands

C   The rules do not limit the number of transmitters that an amateur licensee may control at the same time. [97.5 (d)]

T4B11 If you have been informed that your amateur radio station causes interference to nearby radio or television broadcast receivers of good engineering design, what operating restrictions can FCC rules impose on your station?

A. Require that you discontinue operation on frequencies causing interference during certain evening hours and on Sunday morning (local time)
B. Relocate your station or reduce your transmitter's output power
C. Nothing, unless the FCC conducts an investigation of the interference problem and issues a citation
D. Reduce antenna height so as to reduce the area affected by the interference

A   If a spurious emission from your station is causing interference to nearby radio or television receivers of good engineering design, the FCC can impose operating restrictions on your station. These restrictions may include a requirement to discontinue operation on frequencies causing interference during certain evening hours and on Sunday mornings (local time). The FCC Rules say that these restrictions apply between 8 PM and 10:30 PM local time daily and from 10:30 AM to 1 PM on Sunday. [97.121]

T4B12 How could you best keep unauthorized persons from using your amateur station at home?

A. Use a carrier-operated relay in the main power line
B. Use a key-operated on/off switch in the main power line
C. Put a "Danger - High Voltage" sign in the station
D. Put fuses in the main power line

B   It is important to keep any unauthorized persons from using your Amateur Radio station. This is both a safety concern and a requirement of the FCC Rules. You can install a key-operated on/off switch in the main ac-power line to your station equipment. With the switch turned off and the key in your pocket, you will be sure no one can use your station.

T4B13 How could you best keep unauthorized persons from using a mobile amateur station in your car?

A. Disconnect the microphone when you are not using it
B. Put a "do not touch" sign on the radio
C. Turn the radio off when you are not using it
D. Tune the radio to an unused frequency when you are done using it

A   Safety concerns and FCC Rules require that you keep any unauthorized persons from using your Amateur Radio station. In a mobile installation, you can simply disconnect the microphone and remove it when you leave the vehicle. This is an easy way to prevent unauthorized transmissions.

T4C Providing public service; emergency and disaster communications; Distress calling; Emergency drills and communications; Purpose of RACES.

T4C01 If you hear a voice distress signal on a frequency outside of your license privileges, what are you allowed to do to help the station in distress?
- A. You are NOT allowed to help because the frequency of the signal is outside your privileges
- B. You are allowed to help only if you keep your signals within the nearest frequency band of your privileges
- C. You are allowed to help on a frequency outside your privileges only if you use international Morse code
- D. You are allowed to help on a frequency outside your privileges in any way possible

D   If you receive a distress signal, you are also allowed to transmit on any frequency to provide assistance. [97.405 (a)]

T4C02 When may you use your amateur station to transmit an "SOS" or "MAYDAY"?
- A. Never
- B. Only at specific times (at 15 and 30 minutes after the hour)
- C. In a life—or property—threatening emergency
- D. When the National Weather Service has announced a severe weather watch

C   In a life or property-threatening emergency, you may send a distress call on *any* frequency, even outside the amateur bands, if you think doing so will bring help faster. [97.403]

T4C03 If a disaster disrupts normal communication systems in an area where the FCC regulates the amateur service, what kinds of transmissions may stations make?

A. Those that are necessary to meet essential communication needs and facilitate relief actions

B. Those that allow a commercial business to continue to operate in the affected area

C. Those for which material compensation has been paid to the amateur operator for delivery into the affected area

D. Those that are to be used for program production or newsgathering for broadcasting purposes

A The FCC encourages licensed hams to assist in emergency situations. There are even Rules to get around the Rules that apply at all other times! Section 97.401(a) says: "When normal communication systems are overloaded, damaged or disrupted because a disaster has occurred, or is likely to occur ... an amateur station may make transmissions necessary to meet essential communication needs and facilitate relief actions." [97.401 (a)]

T4C04 What information is included in an FCC declaration of a temporary state of communication emergency?

A. A list of organizations authorized to use radio communications in the affected area

B. A list of amateur frequency bands to be used in the affected area

C. Any special conditions and special rules to be observed during the emergency

D. An operating schedule for authorized amateur emergency stations

C In the wake of a major disaster, the FCC may suspend or change its Rules to help deal with the immediate problem. Part 97 says that when a disaster disrupts normal communications systems in a particular area, the FCC may declare a temporary state of communication emergency. This declaration will set forth any special conditions or rules to be observed during the emergency. [97.401 (c)]

T4C05 If you are in contact with another station and you hear an emergency call for help on your frequency, what should you do?

A. Tell the calling station that the frequency is in use
B. Direct the calling station to the nearest emergency net frequency
C. Call your local Civil Preparedness Office and inform them of the emergency
D. Stop your QSO immediately and take the emergency call

D   Here's what you should do if you're in contact with another station and hear an emergency call for help on your frequency. You should immediately stop your QSO and take the call!

T4C06 What is the proper way to interrupt a repeater conversation to signal a distress call?

A. Say "BREAK" once, then your call sign
B. Say "HELP" as many times as it takes to get someone to answer
C. Say "SOS," then your call sign
D. Say "EMERGENCY" three times

A   To make a distress call during an ongoing repeater conversation, the proper procedure is to say "BREAK" once, followed by your call sign. (On some repeaters, operators may say "BREAK" when they want to join a conversation. In general, you should try to avoid using that practice.) If you hear someone say "BREAK" you should immediately answer them and stand by for their emergency communications.

T4C07 What is one reason for using tactical call signs such as "command post" or "weather center" during an emergency?

A. They keep the general public informed about what is going on
B. They are more efficient and help coordinate public-service communications
C. They are required by the FCC
D. They increase goodwill between amateurs

B   Tactical call signs describe a function, location or agency. Their use promotes coordination with individuals or agencies that are monitoring. When operators change shifts or locations, the set of tactical call signs remains the same. Amateurs may use such tactical call signs as *parade headquarters, finish line, Command Post, Weather Center* or *Net Control*. This procedure promotes efficiency and coordination in public-service communication activities. Tactical call signs do not fulfill the identification requirements of Section 97.119 of the FCC rules, however. Amateurs must also identify their station operation with their FCC-assigned call sign.

T4C08 What type of messages concerning a person's well being are sent into or out of a disaster area?

A. Routine traffic
B. Tactical traffic
C. Formal message traffic
D. Health and welfare traffic

D   Health-and-Welfare traffic pertains to the well being of people in a disaster area. Friends and relatives of those who may have been injured or evacuated want to know if their loved ones are okay. Health-and-Welfare traffic provides information to those waiting outside the disaster area.

T4C09 What are messages called that are sent into or out of a disaster area concerning the immediate safety of human life?

A. Tactical traffic
B. Emergency traffic
C. Formal message traffic
D. Health and welfare traffic

B There can be a large number of messages to handle during a disaster. Phone lines still in working order are often overloaded. "Emergency traffic" messages have life-and-death urgency or are for medical help and critical supplies involving the immediate safety of human life.

T4C10 Why is it a good idea to have a way to operate your amateur station without using commercial AC power lines?

A. So you may use your station while mobile
B. So you may provide communications in an emergency
C. So you may operate in contests where AC power is not allowed
D. So you will comply with the FCC rules

B When the call for emergency communications comes, amateurs answer. Many hams have some way to operate their station without using commercial ac power. Power lines are often knocked down leaving areas without power during a natural disaster such as a hurricane, tornado, earthquake or ice storm.

T4C11 What is the most important accessory to have for a hand-held radio in an emergency?

A. An extra antenna
B. A portable amplifier
C. Several sets of charged batteries
D. A microphone headset for hands-free operation

C One of the most important accessories you can have for your hand-held radio is several extra battery packs. Just be sure you keep them charged. Many hams have found that a battery pack that holds regular alkaline batteries is also an excellent accessory. Alkaline batteries are widely available, they have a long shelf life and will last longer than a single charge in a rechargeable battery pack.

T4C12 Which type of antenna would be a good choice as part of a portable HF amateur station that could be set up in case of an emergency?

A. A three-element quad
B. A three-element Yagi
C. A dipole
D. A parabolic dish

C A dipole antenna is an excellent choice for a portable HF station that can be set up in emergencies. It can be installed easily, and wire is light and portable. Carry an ample supply of wire and you'll be ready to go on the air at any time and any place.

T4C13 How must you identify messages sent during a RACES drill?

A. As emergency messages
B. As amateur traffic
C. As official government messages
D. As drill or test messages

D Only civil-preparedness communications can be transmitted during RACES operation. These are defined in section 97.407 of the FCC regulations. Rules permit tests and drills for a maximum of one hour per week. All test and drill messages must be clearly identified as such.

T4C14 With what organization must you register before you can participate in RACES drills?

A. A local Amateur Radio club
B. A local racing organization
C. The responsible civil defense organization
D. The Federal Communications Commission

C You must be registered with the responsible civil defense organization to operate as a RACES station or participate in RACES drills. RACES stations may not communicate with amateurs not operating in a RACES capacity.

Control Operator Duties

There will be 3 questions on your exam taken from the Control Operator Duties subelement printed in this chapter. These questions are divided into 3 groups, labeled T5A, T5B and T5C.

After some of the explanations in this chapter you will see a reference to Part 97 of the Federal Communications Commission Rules, set inside square brackets, like [97.3 (a) (5)]. This tells you where to look for the exact wording of the Rules as they relate to that question. For a complete copy of Part 97, along with simple explanations of the Rules governing Amateur Radio, see *The ARRL's FCC Rule Book*.

T5A   Determining operating privileges, Where control operator must be situated while station is locally or remotely controlled; Operating other amateur stations.

T5A01 If you are the control operator at the station of another amateur who has a higher-class license than yours, what operating privileges are you allowed?

A. Any privileges allowed by the higher license
B. Only the privileges allowed by your license
C. All the emission privileges of the higher license, but only the frequency privileges of your license
D. All the frequency privileges of the higher license, but only the emission privileges of your license

B   Suppose that you hold a Technician class license and are the control operator at the station of another amateur with a higher-class license than yours (a General, Advanced or Amateur Extra class.) In this situation, you can use only the privileges allowed by your license! [97.105 (b)]

T5A02 Assuming you operate within your amateur license privileges, what restrictions apply to operating amateur equipment?

A. You may operate any amateur equipment

B. You may only operate equipment located at the address printed on your amateur license

C. You may only operate someone else's equipment if you first notify the FCC

D. You may only operate store-purchased equipment until you earn your Amateur Extra class license

A   Your Technician license authorizes you to be the control operator of an amateur station in the Technician frequency bands. This means you can be the control operator of your own station or someone else's station. In either case, you are responsible to the FCC for the proper operation of the station. You may operate any amateur equipment provided that you do so within your amateur license privileges.

T5A03 When an amateur station is transmitting, where must its control operator be, assuming the station is not under automatic control?

A. At the station's control point

B. Anywhere in the same building as the transmitter

C. At the station's entrance, to control entry to the room

D. Anywhere within 50 km of the station location

A   Unless a station is operating under automatic control, a control operator must be present at the station control point whenever a transmitter is operating. The control point is simply the location where the control operator function is performed. [97.109 (b)]

T5A04 Where will you find a detailed list of your operating privileges?

A. In the OET Bulletin 65 Index

B. In FCC Part 97

C. In your equipment's operating instructions

D. In Part 15 of the Code of Federal Regulations

B   You will find a detailed list of your operating privileges in Part 97 of the FCC rules.

T5A05 If you transmit from another amateur's station, who is responsible for its proper operation?

A. Both of you

B. The other amateur (the station licensee)

C. You, the control operator

D. The station licensee, unless the station records show that you were the control operator at the time

A   Any amateur may designate another licensed operator as the control operator. However, the FCC holds both the control operator and the station licensee responsible for proper operation of the station. [97.103 (a)]

T5A06 If you let another amateur with a higher class license than yours control your station, what operating privileges are allowed?

A. Any privileges allowed by the higher license, as long as proper identification procedures are followed

B. Only the privileges allowed by your license

C. All the emission privileges of the higher license, but only the frequency privileges of your license

D. All the frequency privileges of the higher license, but only the emission privileges of your license

A   If you let another amateur with a higher-class license than yours control your station, any privileges allowed by his or her license are permitted. They're permitted as long as proper identification procedures are followed. [97.105 (b)]

T5A07 If a Technician class licensee uses the station of a General class licensee, how may the Technician licensee operate?

A. Within the frequency limits of a General class license

B. Within the limits of a Technician class license

C. Only as a third party with the General class licensee as the control operator

D. A Technician class licensee may not operate a General class station

B   You can hold a Technician class license and be the control operator at the station of another amateur with a higher-class license than yours (a General, Advanced or Amateur Extra class.) However, you can use only the privileges allowed by your license. [97.105 (b)]

T5A08 What type of amateur station does not require the control operator to be present at the control point?

    A. A locally controlled station
    B. A remotely controlled station
    C. An automatically controlled station
    D. An earth station controlling a space station

C   A station operating under automatic control does not require the control operator to be present at the control point. Automatic control is defined by the FCC as "The use of devices and procedures for control of a station when it is transmitting so that compliance with the FCC Rules is achieved without the control operator being present at a control point." This is "hands off" operation; there's nobody at the control point, *but there must still be a control operator available who is responsible*. Only stations specifically designated in Part 97 may be automatically controlled. [97.109 (d)]

T5A09 Why can't unlicensed persons in your family transmit using your amateur station if they are alone with your equipment?

    A. They must not use your equipment without your permission
    B. They must be licensed before they are allowed to be control operators
    C. They must first know how to use the right abbreviations and Q signals
    D. They must first know the right frequencies and emissions for transmitting

B   You may not allow an unlicensed person—even a family member—to operate your radio transmitter while you are not present. They must first be licensed before they're allowed to be control operators. [97.109 (b)]

T5A10 If you own a dual-band mobile transceiver, what requirement must be met if you set it up to operate as a crossband repeater?

    A. There is no special requirement if you are licensed for both bands
    B. You must hold an Amateur Extra class license
    C. There must be a control operator at the system's control point
    D. Operating a crossband mobile system is not allowed

C   A repeater station is an amateur station that automatically retransmits the signals of other stations. In that type of operation there is typically no control operator at the transmitter control point, but there is some way to turn off the repeater if there is a problem. What if you wanted to set up your dual-band mobile transceiver as a crossband repeater? In this instance, a control operator is required at the system's control point because the mobile radio has no way to turn it off without a control operator at the controls.

T5B01 How often must an amateur station be identified?

A. At the beginning of a contact and at least every ten minutes after that
B. At least once during each transmission
C. At least every ten minutes during and at the end of a contact
D. At the beginning and end of each transmission

C   FCC regulations are very specific about station identification. You must identify your station every ten minutes (or more frequently) during a contact and at the end of the contact. [97.119 (a)]

T5B02 What identification, if any, is required when two amateur stations end communications?

A. No identification is required
B. One of the stations must transmit both stations' call signs
C. Each station must transmit its own call sign
D. Both stations must transmit both call signs

C   You do not have to transmit both call signs when you are talking with another ham—only your own. Nothing prevents you from giving both call signs, but you're only required to transmit you own call sign. [97.119 (a)]

T5B03 What is the longest period of time an amateur station can operate without transmitting its call sign?

A. 5 minutes
B. 10 minutes
C. 15 minutes
D. 30 minutes

B   It doesn't matter if you are operating from your home station, from a portable location or in a vehicle as a mobile station. It doesn't matter what mode you are operating. You must always identify your station every ten minutes when you are operating. [97.119 (a)]

T5B04 This question has been withdrawn.

T5B05 What is the term for the average power supplied to an antenna transmission line during one RF cycle at the crest of the modulation envelope?

A. Peak transmitter power
B. Peak output power
C. Average radio-frequency power
D. Peak envelope power

D   The FCC defines peak envelope power as "the average power supplied to the antenna transmission line by a transmitter during one RF cycle at the crest of the modulation envelope." This sounds pretty technical, but it isn't too difficult to understand the basic principle. The modulation envelope refers to the way the information signal varies the transmitter output. Think of it as increasing and decreasing the transmitted signal. All you have to do is find the highest point, or maximum output-signal level. Then you look at one cycle of the radio-frequency (RF) signal and measure the average power during that time. [97.3 (b) (6)]

T5B06 This question has been withdrawn.

T5B07 What amount of transmitter power must amateur stations use at all times?

A. 25 watts PEP output
B. 250 watts PEP output
C. 1500 watts PEP output
D. The minimum legal power necessary to communicate

D   According to FCC Rules, an amateur station must use the minimum transmitter power necessary to maintain reliable communication. What this means is simple—if you don't need 200 W to contact someone, don't use it! For example, suppose you contact another amateur station and learn that your signals are extremely strong and loud, and perfectly readable. You should turn down your transmitter output power in that case. [97.313 (a)]

T5B08 If you are using a language besides English to make a contact, what language must you use when identifying your station?

A. The language being used for the contact
B. The language being used for the contact, provided the US has a third-party communications agreement with that country
C. English
D. Any language of a country that is a member of the International Telecommunication Union

C   You can use any language you want to communicate with other amateurs. Amateur Radio provides you a great opportunity to practice your foreign language skills with other hams that speak that language. When you give your station identification, however, you must use English. [97.119 (b) (2)]

T5B09 If you are helping in a communications emergency that is being handled by a net control operator, how might you best minimize interference to the net once you have checked in?

A. Whenever the net frequency is quiet, announce your call sign and location
B. Move 5 kHz away from the net's frequency and use high power to ask for other emergency communications
C. Do not transmit on the net frequency until asked to do so by the net operator
D. Wait until the net frequency is quiet, then ask for any emergency traffic for your area

C   Suppose you are helping in a communications emergency that is being handled by a net control operator. How can you best minimize interference to the net once you have checked in? The simple but often overlooked answer is to not transmit on the net frequency until asked to do so by the net control operator. This act of self-restraint will help prevent needless confusion in a situation where efficiency matters.

T5B10 What are the station identification requirements for an amateur transmitter used for telecommand (control) of model craft?

A. Once every ten minutes
B. Once every ten minutes, and at the beginning and end of each transmission
C. At the beginning and end of each transmission
D. Station identification is not required if the transmitter is labeled with the station licensee's name, address and call sign

D   There is an exception to the station identification rules when you are transmitting signals to control a model craft, such as a plane or boat. Such control or telecommand operation is simply a one-way transmission to start, change or end functions of a device at a distance. The model craft's control transmitter must be labeled, however. The label must contain the station licensee's name, address and call sign. Also, the model craft's control transmitter power cannot exceed one watt. [97.215 (a)]

T5B11 Why is transmitting on a police frequency as a "joke" called harmful interference that deserves a large penalty?

    A. It annoys everyone who listens

    B. It blocks police calls that might be an emergency and interrupts police communications

    C. It is in bad taste to communicate with non-amateurs, even as a joke

    D. It is poor amateur practice to transmit outside the amateur bands

    B   While FCC rules permit a broad standard of operating flexibility for emergency communications, they must not be taken lightly. Transmitting on a police frequency as a joke, for example, is harmful interference. A large penalty will almost certainly be the result of such an act! Such interference can block police calls that might be an emergency, and interrupt police communications. [97.3 (a) (23)]

T5B12 If you are using a frequency within a band assigned to the amateur service on a secondary basis, and a station assigned to the primary service on that band causes interference, what action should you take?

    A. Notify the FCC's regional Engineer in Charge of the interference

    B. Increase your transmitter's power to overcome the interference

    C. Attempt to contact the station and request that it stop the interference

    D. Change frequencies; you may be causing harmful interference to the other station, in violation of FCC rules

    D   Say you are using a frequency within a band assigned to the amateur service on a secondary basis. A station assigned to the primary service on that same band begins to cause interference to your operation. What action should you take? As an operator in a secondary service, *YOU* should change frequencies! You may be causing harmful interference to the other station with primary service status, which is a violation of FCC rules. [97.303]

T5C Authorized transmissions, Prohibited practices; Third party communications; Retransmitting radio signals; One way communications.

T5C01 If you answer someone on the air and then complete your communication without giving your call sign, what type of communication have you just conducted?

A. Test transmission
B. Tactical signal
C. Packet communication
D. Unidentified communication

D   You should not press the push-to-talk button on your radio or microphone to test your repeater access without giving your call sign. What if you answer someone on the air and complete your communication without giving your call sign? These are both examples of unidentified communications, and they're illegal. [97.119 (a)]

T5C02 What is one example of one-way communication that Technician class control operators are permitted by FCC rules?

A. Transmission for radio control of model craft
B. Use of amateur television for surveillance purposes
C. Retransmitting National Weather Service broadcasts
D. Use of amateur radio as a wireless microphone for a public address system

A   Emergency communications, remote control of a model craft and beacon operation are all examples of permitted one-way communications. These are transmissions that are not intended to be answered. The FCC strictly limits the types of one-way communications allowed on the amateur bands. [97.111 (b) (3)]

T5C03 What kind of payment is allowed for third-party messages sent by an amateur station?

A. Any amount agreed upon in advance
B. Donation of repairs to amateur equipment
C. Donation of amateur equipment
D. No payment of any kind is allowed

D   The Amateur Service is not the place to conduct business. This means that you may not receive any type of payment in return for transmitting or receiving third-party communication. It is expressly forbidden by the FCC Rules. [97.11 (a) (2)]

T5C04 What is the definition of third-party communications?

A. A message sent between two amateur stations for someone else
B. Public service communications for a political party
C. Any messages sent by amateur stations
D. A three-minute transmission to another amateur

A   A message sent between two amateur stations for someone else is third-party communications. (Many hams call it third-party traffic.) For example, sending a message from your mother-in-law to her relatives in Scarsdale on Valentine's Day is third-party communications. [97.3 (a) (44)]

T5C05 When are third-party messages allowed to be sent to a foreign country?

A. When sent by agreement of both control operators
B. When the third party speaks to a relative
C. They are not allowed under any circumstances
D. When the US has a third-party agreement with the foreign country or the third party is qualified to be a control operator

D   FCC Rules strictly limit international third-party communications to those with countries that have third-party communications agreements with the US. These agreements allow amateurs in both countries to participate in third-party communications. You may also pass these messages when the third party is eligible to be a control operator of the station. [97.115 (a) (2)]

T5C06 If you let an unlicensed third party use your amateur station, what must you do at your station's control point?

A. You must continuously monitor and supervise the third-party's participation
B. You must monitor and supervise the communication only if contacts are made in countries that have no third-party communications agreement with the US
C. You must monitor and supervise the communication only if contacts are made on frequencies below 30 MHz
D. You must key the transmitter and make the station identification

A   You may allow an unlicensed person to participate in Amateur Radio from your station. This is third-party participation. It is another form of third-party communications. You (as control operator) must always be present to make sure the unlicensed person follows all the rules. [97.115 (b) (1)]

T5C07 Besides normal identification, what else must a US station do when sending third-party communications internationally?

A. The US station must transmit its own call sign at the beginning of each communication, and at least every ten minutes after that
B. The US station must transmit both call signs at the end of each communication
C. The US station must transmit its own call sign at the beginning of each communication, and at least every five minutes after that
D. Each station must transmit its own call sign at the end of each transmission, and at least every five minutes after that

B   In normal operation, you are only required to transmit your own call sign. However when you are exchanging international third-party communications, you must transmit both call signs at the end of each communication. [97.115 (c)]

T5C08 If an amateur pretends there is an emergency and transmits the word "MAYDAY," what is this called?

A. A traditional greeting in May
B. An emergency test transmission
C. False or deceptive signals
D. Nothing special; "MAYDAY" has no meaning in an emergency

C   FCC rules prohibit the transmission of false or deceptive signals. These are transmissions intended to mislead or confuse those who receive them. As an example would be if someone were to pretend there is an emergency and transmit MAYDAY! [97.113 (a) (4)]

T5C09 If an amateur transmits to test access to a repeater without giving any station identification, what type of communication is this called?

A. A test emission; no identification is required
B. An illegal unmodulated transmission
C. An illegal unidentified transmission
D. A non-communication; no voice is transmitted

C   The rules prohibit unidentified communications or signals. These are signals where the transmitting station's call sign is not included. Be sure you understand the proper station identification procedures, so you don't violate this rule. You should not press the push-to-talk button on your radio or microphone to test your repeater access without giving your call sign. [97.119 (a)]

T5C10 When may you deliberately interfere with another station's communications?

A. Only if the station is operating illegally
B. Only if the station begins transmitting on a frequency you are using
C. Never
D. You may expect, and cause, deliberate interference because it can't be helped during crowded band conditions

C   Another prohibited practice is deliberately interfering with another station's communications. For example, you must not repeatedly transmit on a frequency that is already occupied. It could be over a group of amateurs in a net, or just two hams already in a QSO. Regardless, each is an example of harmful or malicious interference. [97.101 (d)]

T5C11 If an amateur repeatedly transmits on a frequency already occupied by a group of amateurs in a net operation, what type of interference is this called?

A. Break-in interference
B. Harmful or malicious interference
C. Incidental interference
D. Intermittent interference

B   This deliberate act is an example of harmful or malicious interference. It is illegal and should be avoided. [97.3 (a) (22)]

T5C12 What device is commonly used to retransmit amateur radio signals?

A. A beacon
B. A repeater
C. A radio controller
D. A duplexer

B   A repeater station is an amateur station that automatically retransmits the signals of other stations.

Good Operating Practices

There will be 3 questions on your exam taken from the Good Operating Practices subelement printed in this chapter. These questions are divided into 3 groups, labeled T6A, T6B and T6C.

After some of the explanations in this chapter you will see a reference to Part 97 of the Federal Communications Commission Rules, set inside square brackets, like [97.3 (a) (5)]. This tells you where to look for the exact wording of the Rules as they relate to that question. For a complete copy of Part 97, along with simple explanations of the Rules governing Amateur Radio, see *The ARRL's FCC Rule Book.*

T6A Calling another station; Calling CQ; Typical amateur service radio contacts; Courtesy and respect for others; Popular Q-signals; Signal reception reports; Phonetic alphabet for voice operations.

T6A01 What is the advantage of using the International Telecommunication Union (ITU) phonetic alphabet when identifying your station?

A. The words are internationally recognized substitutes for letters
B. There is no advantage
C. The words have been chosen to represent Amateur Radio terms
D. It preserves traditions begun in the early days of Amateur Radio

A   If the other operator is having difficulty copying your signals you should use the standard International Telecommunication Union (ITU) phonetic alphabet, detailed in **Table 6-1**. Use the words in the phonetic alphabet to spell out the letters in your call sign, your name or any other piece of information that might be confused if the letters are not received correctly. This phonetic alphabet is generally understood by hams in all countries. [97.119 (b) (2)]

Table 6-1
Standard ITU Phonetics

Letter	Word	Pronunciation
A	Alfa	**AL** FAH
B	Bravo	**BRAH** VOH
C	Charlie	**CHAR** LEE
D	Delta	**DELL** TAH
E	Echo	**ECK** OH
F	Foxtrot	**FOKS** TROT
G	Golf	GOLF
H	Hotel	HOH **TELL**
I	India	**IN** DEE AH
J	Juliett	**JEW** LEE ETT
K	Kilo	**KEY** LOH
L	Lima	**LEE** MAH
M	Mike	MIKE
N	November	NO **VEM** BER
O	Oscar	**OSS** CAH
P	Papa	PAH **PAH**
Q	Quebec	KEH **BECK**
R	Romeo	**ROW** ME OH
S	Sierra	SEE **AIR** RAH
T	Tango	**TANG** GO
U	Uniform	**YOU** NEE FORM
V	Victor	**VIK** TAH
W	Whiskey	**WISS** KEY
X	X-Ray	**ECKS** RAY
Y	Yankee	**YANG** KEY
Z	Zulu	**ZOO** LOO

Note: The **boldfaced** syllables are emphasized. The pronunciations shown in this table were designed for those who speak any of the international languages. The pronunciations given for "Oscar" and "Victor" may seem awkward to English-speaking people in the US.

T6A02 What is one reason to avoid using "cute" phrases or word combinations to identify your station?

 A. They are not easily understood by non-English-speaking amateurs
 B. They might offend English-speaking amateurs
 C. They do not meet FCC identification requirements
 D. They might be interpreted as codes or ciphers intended to obscure the meaning of your identification

A Whether you're working a DX operator who may not fully understand our language, or talking to your friend down the street, avoid using cute phrases or word combinations to identify your station. These can be confusing to anybody, and they are not easily understood by non-English-speaking amateurs. [97.119 (b) (2)]

T6A03 **What should you do before you transmit on any frequency?**

A. Listen to make sure others are not using the frequency
B. Listen to make sure that someone will be able to hear you
C. Check your antenna for resonance at the selected frequency
D. Make sure the SWR on your antenna feed line is high enough

A   The first rule of good operating practice is to always listen before you transmit! This may seem so obvious that it's not worth a mention. However, a few seconds of listening will help ensure that you don't interfere with a conversation (QSO) in progress.

T6A04 **How do you call another station on a repeater if you know the station's call sign?**

A. Say "break, break 79," then say the station's call sign
B. Say the station's call sign, then identify your own station
C. Say "CQ" three times, then say the station's call sign
D. Wait for the station to call "CQ," then answer it

B   To call another station when the repeater is not in use, just give both calls. For example, "N1II, this is N1BKE." If the repeater is in use, but the conversation sounds like it is about to end, wait before calling another station.

T6A05 **What does RST mean in a signal report?**

A. Recovery, signal strength, tempo
B. Recovery, signal speed, tone
C. Readability, signal speed, tempo
D. Readability, signal strength, tone

D   RST refers to the readability, strength and tone system of reporting signal reception. You'll exchange a signal report in nearly every Amateur Radio QSO. The scales are simply a general indication of how you are receiving the other station. As you gain experience with the descriptions given in **Table 6-2**, you'll be more comfortable estimating the appropriate signal report.

Table 6-2

The RST System

READABILITY
1—Unreadable.
2—Barely readable, occasional words distinguishable.
3—Readable with considerable difficulty.
4—Readable with practically no difficulty.
5—Perfectly readable.

SIGNAL STRENGTH
1—Faint signals barely perceptible.
2—Very weak signals.
3—Weak signals.
4—Fair signals.
5—Fairly good signals.
6—Good signals.
7—Moderately strong signals.
8—Strong signals.
9—Extremely strong signals.

TONE
1—Sixty-cycle ac or less, very rough and broad.
2—Very rough ac, very harsh and broad.
3—Rough ac tone, rectified but not filtered.
4—Rough note, some trace of filtering.
5—Filtered rectified ac but strongly ripple-modulated.
6—Filtered tone, definite trace of ripple modulation.
7—Near pure tone, trace of ripple modulation.
8—Near perfect tone, slight trace of modulation.
9—Perfect tone, no trace of ripple or modulation of any kind.

The "tone" report refers only to the purity of the signal. It has no connection with its stability or freedom from clicks or chirps. Most of the signals you hear will be a T-9. Other tone reports occur mainly if the power supply filter capacitors are not doing a thorough job. If so, some trace of ac ripple finds its way onto the transmitted signal. If the signal has the characteristic steadiness of crystal control, add X to the report (for example, RST 469X). If it has a chirp or "tail" (either on "make" or "break") add C (for example, 469C). If it has clicks or noticeable other keying transients, add K (for example, 469K). Of course a signal could have both chirps and clicks, in which case both C and K could be used (for example, RST 469CK).

T6A06 What is the meaning of: "Your signal report is five nine plus 20 dB . . . "?

A. Your signal strength has increased by a factor of 100
B. Repeat your transmission on a frequency 20 kHz higher
C. The bandwidth of your signal is 20 decibels above linearity
D. A relative signal-strength meter reading is 20 decibels greater than strength 9

D   In Table 6-2 you'll see that the "five" means your signal is perfectly readable. The "nine" means that your signal is extremely strong. The "plus 20 dB" would mean that your signal is 20 dB stronger than "extremely strong." As a practical matter, this signal report usually means that the other station is getting a reading 20 decibels greater than strength 9 on the receiver's S meter.

T6A07 What is the meaning of the procedural signal "CQ"?

A. Call on the quarter hour
B. New antenna is being tested (no station should answer)
C. Only the called station should transmit
D. Calling any station

D   CQ literally means "Seek you: Calling any station." You can usually tell a good ham by the length of the CQ call. A good operator sends short calls that are separated by listening periods. Long CQs drive away more contacts than they attract!

T6A08 What is a QSL card in the amateur service?

A. A letter or postcard from an amateur pen pal
B. A Notice of Violation from the FCC
C. A written acknowledgment of communications between two amateurs
D. A postcard reminding you when your license will expire

C   QSL means, "I acknowledge receipt" (of a message or information). A QSL card is a written confirmation of an amateur contact (QSO).

T6A09 What is the correct way to call CQ when using voice?

    A. Say "CQ" once, followed by "this is," followed by your call sign spoken three times

    B. Say "CQ" at least five times, followed by "this is," followed by your call sign spoken once

    C. Say "CQ" three times, followed by "this is," followed by your call sign spoken three times

    D. Say "CQ" at least ten times, followed by "this is," followed by your call sign spoken once

    C   Keep your CQ calls short. The format of a correct CQ call on voice is a "3 × 3" call. This refers to calling CQ three times, then giving your call sign three times. Here's an example of a CQ call: "CQ CQ Calling CQ. This is KB1AFE, Kilo Bravo One Alfa Foxtrot Echo, KB1AFE calling CQ and standing by."

T6A10 How should you answer a voice CQ call?

    A. Say the other station's call sign at least ten times, followed by "this is," then your call sign at least twice

    B. Say the other station's call sign at least five times phonetically, followed by "this is," then your call sign at least once

    C. Say the other station's call sign at least three times, followed by "this is," then your call sign at least five times phonetically

    D. Say the other station's call sign once, followed by "this is," then your call sign given phonetically

    D   When replying to a CQ, say both call signs clearly. It's not necessary to sign the other station's call phonetically. You should always sign yours with standard phonetics, however. Remember to keep calls short.

T6A11 What is the meaning of: "Your signal is full quieting..."?

    A. Your signal is strong enough to overcome all receiver noise

    B. Your signal has no spurious sounds

    C. Your signal is not strong enough to be received

    D. Your signal is being received, but no audio is being heard

    A   On FM repeaters, RS reports are not used. FM signal reports are generally given in terms of signal quieting. Full quieting means the received signal is strong enough to overcome all receiver noise.

T6A12 What is meant by the term "DX"?

A. Best regards
B. Distant station
C. Calling any station
D. Go ahead

B DX means any distant station. Most operators understand DX to mean a station in another country, no matter how near or far that may be. In fact, it's a relative term. On the VHF/UHF bands, where one normally expects the contacts to be limited to one's local area, a station 100 miles away might become DX.

T6A13 What is the meaning of the term "73"?

A. Long distance
B. Best regards
C. Love and kisses
D. Go ahead

B 73 is a common abbreviation that means "Best regards." Most hams say "73" at the end of a phone contact, too. This is one of those CW abbreviations that has made its way into the mainstream of Amateur Radio lingo.

T6B   Occupied bandwidth for emission types; Mandated and voluntary band plans; CW operation.

T6B01 Which list of emission types is in order from the narrowest bandwidth to the widest bandwidth?

A. RTTY, CW, SSB voice, FM voice
B. CW, FM voice, RTTY, SSB voice
C. CW, RTTY, SSB voice, FM voice
D. CW, SSB voice, RTTY, FM voice

C An FM voice signal occupies more than twice the bandwidth of a comparable SSB voice signal. CW has a narrower bandwidth than an RTTY signal. The correct order from the narrowest bandwidth to the widest bandwidth is CW, RTTY, SSB voice, and FM voice. See the drawing on the next page.

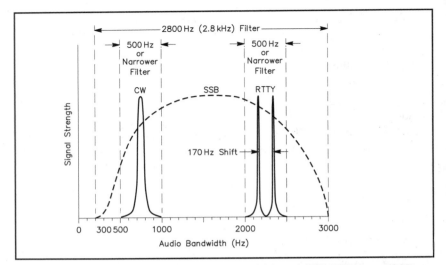

This drawing illustrates the relative bandwidths of CW, RTTY and SSB signals. The bandwidths of filters that might be used in a receiver's intermediate frequency (IF) section to receive these signals are also shown.

T6B02 What is the usual bandwidth of a single-sideband amateur signal?

    A. 1 kHz
    B. 2 kHz
    C. Between 3 and 6 kHz
    D. Between 2 and 3 kHz

    D   SSB usually has a bandwidth between 2 and 3 kHz. Your voice contains frequencies higher than 3 kHz, but all of the sounds necessary to understand speech are between about 300 Hz and 3000 Hz. Most amateur voice transmitters limit the bandwidth of a transmitted audio signal to between 300 and 3000 Hz. The difference between these limits is the bandwidth, 2700 Hz.

T6B03 What is the usual bandwidth of a frequency-modulated amateur signal?

    A. Less than 5 kHz
    B. Between 5 and 10 kHz
    C. Between 10 and 20 kHz
    D. Greater than 20 kHz

    C   An FM transmitter using 5-kHz deviation and a maximum audio frequency of 3 kHz uses a total bandwidth of about 16 kHz. The actual bandwidth of a typical FM signal may be somewhat greater than this. A good approximation is that the bandwidth of an FM voice signal is between 10 and 20 kHz.

T6B04 What is the usual bandwidth of a UHF amateur fast-scan television signal?

A. More than 6 MHz
B. About 6 MHz
C. About 3 MHz
D. About 1 MHz

B   The usual bandwidth of a UHF amateur fast-scan television signal is about 6 MHz. You can see the details in the drawing.

A 6-MHz ATV video channel with the video carrier 1.25 MHz up from the lower edge. The color subcarrier is at 3.58 MHz and the sound subcarrier at 4.5 MHz above the video carrier.

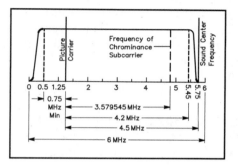

T6B05 What name is given to an amateur radio station that is used to connect other amateur stations with the Internet?

A. A gateway
B. A repeater
C. A digipeater
D. FCC regulations prohibit such a station

A   A gateway is a special type of amateur station. Its purpose is to interface between radio communications links and the Internet. In other words, a gateway is used to connect other amateur stations with the Internet.

T6B06 What is a band plan?

A. A voluntary guideline beyond the divisions established by the FCC for using different operating modes within an amateur band
B. A guideline from the FCC for making amateur frequency band allocations
C. A plan of operating schedules within an amateur band published by the FCC
D. A plan devised by a club to best use a frequency band during a contest

A   Band plans are voluntary agreements between operators about how to use the bands. These plans go into more detail on using different operating modes within an amateur band than specified in FCC regulations. Good operators are familiar with the band plans and try to follow them.

T6B07 At what speed should a Morse code CQ call be transmitted?

A. Only speeds below five WPM
B. The highest speed your keyer will operate
C. Any speed at which you can reliably receive
D. The highest speed at which you can control the keyer

C   Make sure to send Morse code at a speed you can reliably copy. Sending too fast will surely invite disaster! This is especially true when you call CQ. If you answer a CQ, answer at a speed no faster than that of the sending station. Don't be ashamed to send PSE QRS (please send more slowly) if you are having trouble copying at the other stations sending speed.

T6B08 What is the meaning of the procedural signal "DE"?

A. "From" or "this is," as in "WØAIH DE KA9FOX"
B. "Directional Emissions" from your antenna
C. "Received all correctly"
D. "Calling any station"

A   DE is a Morse code abbreviation that means "from" or "this is." When operating Morse code, you use DE before your call sign to clearly identify that the transmission is from your station.

T6B09 What is a good way to call CQ when using Morse code?

A. Send the letters "CQ" three times, followed by "DE," followed by your call sign sent once

B. Send the letters "CQ" three times, followed by "DE," followed by your call sign sent three times

C. Send the letters "CQ" ten times, followed by "DE," followed by your call sign sent twice

D. Send the letters "CQ" over and over until a station answers

B   Morse code CQ calls should follow the same general procedure as an SSB CQ. Generally, a "3 × 3" call is sufficient. Here's an example: CQ CQ CQ DE KA7XYZ KA7XYZ KA7XYZ K. As you may remember, "3 × 3" refers to calling CQ three times, followed by your call sign three times.

T6B10 How should you answer a Morse code CQ call?

A. Send your call sign four times

B. Send the other station's call sign twice, followed by "DE," followed by your call sign twice

C. Send the other station's call sign once, followed by "DE," followed by your call sign four times

D. Send your call sign followed by your name, station location and a signal report

B   If you are sending at an appropriate speed, (no faster than the station calling CQ) you shouldn't have to repeat a lot. Your goal is to get the other station's attention and let the operator know that you are calling. Of course, you'll have to transmit your call sign—a couple of times should be enough. A typical answer to a Morse code CQ is to send the other station's call sign twice, followed by "DE," followed by your call sign twice.

T6B11 What is the meaning of the procedural signal "K"?

A. "Any station transmit"

B. "All received correctly"

C. "End of message"

D. "Called station only transmit"

A   K is a Morse code abbreviation that means, "Any station transmit." It indicates that you have finished transmitting and are now going to listen.

T6B12 What is one meaning of the Q signal "QRS"?

A. "Interference from static"
B. "Send more slowly"
C. "Send RST report"
D. "Radio station location is"

B   QRS is a Morse code abbreviation that means, "Send more slowly." This simple abbreviation is a great friend to hams who are just starting to operate CW. You should not hesitate to send QRS if you are having trouble copying the other fellow's code. Also, if someone sends QRS to you, be sure to slow your speed.

T6C   TVI and RFI reduction and elimination, Band/Low/High pass filter, Out of band harmonic Signals, Spurious Emissions, Telephone Interference, Shielding, Receiver Overload.

T6C01 What is meant by receiver overload?

A. Too much voltage from the power supply
B. Too much current from the power supply
C. Interference caused by strong signals from a nearby source
D. Interference caused by turning the volume up too high

C   Receiver overload is a common type of TV and FM-broadcast interference. It happens most often to consumer electronic equipment near an amateur station or other transmitter. When the RF signal (at the fundamental frequency) enters the receiver, it overloads one or more circuits. The receiver front end (first RF amplifier stage after the antenna) is most commonly affected.

T6C02 What type of filter might be connected to an amateur HF transmitter to cut down on harmonic radiation?

A. A key-click filter
B. A low-pass filter
C. A high-pass filter
D. A CW filter

B   A low-pass filter allows RF energy in the amateur bands to pass freely. It blocks very high frequency harmonics that can fall in the TV and FM bands. Low-pass filters usually have a specified cutoff frequency, often around 40 MHz, above which they severely attenuate the passage of RF energy.

T6C03 What type of filter should be connected to a TV receiver as the first step in trying to prevent RF overload from an amateur HF station transmission?

- A. Low-pass
- B. High-pass
- C. Band pass
- D. Notch

B RF overload can only be cured at the receiver. The first step is to install a high-pass filter on the TV receiver. A high-pass filter is a tuned circuit that passes high frequencies (TV channels start at 54 MHz). The filter blocks low frequencies (the HF amateur bands are in the range of 1.8-30 MHz).

T6C04 What effect might a break in a cable television transmission line have on amateur communications?

- A. Cable lines are shielded and a break cannot affect amateur communications
- B. Harmonic radiation from the TV receiver may cause the amateur transmitter to transmit off-frequency
- C. TV interference may result when the amateur station is transmitting, or interference may occur to the amateur receiver
- D. The broken cable may pick up very high voltages when the amateur station is transmitting

C Any loose connector or break in the transmission line of a cable TV system can allow amateur signals to "leak" into the line, causing interference to TV receivers. Such a leak in the system can also allow Cable TV signals to leak out of the system and cause interference to amateur receivers using that frequency. Cable TV systems typically use some amateur VHF/UHF frequencies to carry the signals inside the cable. This causes no problems as long as there are no leaks in the system.

T6C05 If you are told that your amateur station is causing television interference, what should you do?

    A. First make sure that your station is operating properly, and that it does not cause interference to your own television

    B. Immediately turn off your transmitter and contact the nearest FCC office for assistance

    C. Connect a high-pass filter to the transmitter output and a low-pass filter to the antenna-input terminals of the television

    D. Continue operating normally, because you have no reason to worry about the interference

    A   If a neighbor complains of television interference, you should first make sure that your equipment is operating properly. Check for interference to your own TV. If you see it, stop operating and cure the problem before you go back on the air.

T6C06 If harmonic radiation from your transmitter is causing interference to television receivers in your neighborhood, who is responsible for taking care of the interference?

    A. The owners of the television receivers are responsible

    B. Both you and the owners of the television receivers share the responsibility

    C. You alone are responsible, since your transmitter is causing the problem

    D. The FCC must decide if you or the owners of the television receivers are responsible

    C   Interference caused by harmonics radiated from your station must be cured at your transmitter. As a licensed amateur, you must take steps to see that harmonics from your transmitter do not interfere with other services. You alone are responsible, since your transmitter is causing the problem.

T6C07 If signals from your transmitter are causing front-end overload in your neighbor's television receiver, who is responsible for taking care of the interference?

    A. You alone are responsible, since your transmitter is causing the problem

    B. Both you and the owner of the television receiver share the responsibility

    C. The FCC must decide if you or the owner of the television receiver are responsible

    D. The owner of the television receiver is responsible

D   You should realize that there is nothing you can do to your transmitter to cure receiver overload. It is a fundamental problem with the receiving system, and the primary responsibility for curing the problem is with the owner. The receiver manufacturer should help, but it is the owner that must take the initiative.

T6C08 What circuit blocks RF energy above and below certain limits?

    A. A band-pass filter

    B. A high-pass filter

    C. An input filter

    D. A low-pass filter

A   A band-pass filter is a combination of a high-pass and low-pass filter. It passes a desired range of frequencies while rejecting signals above and below the pass band.

T6C09 If someone tells you that signals from your hand-held transceiver are interfering with other signals on a frequency near yours, what may be the cause?

    A. You may need a power amplifier for your hand-held

    B. Your hand-held may have chirp from weak batteries

    C. You may need to turn the volume up on your hand-held

    D. Your hand-held may be transmitting spurious emissions

D   An ideal transmitter emits a signal only on the operating frequency and nowhere else. Real-world transmitters may also radiate undesired signals, or spurious emissions. Any transmitter can produce spurious emissions: it doesn't matter if you are using a 100-watt HF transceiver or a 2-watt hand-held VHF or UHF radio.

T6C10 What may happen if an SSB transmitter is operated with the microphone gain set too high?

- A. It may cause digital interference to computer equipment
- B. It may cause splatter interference to other stations operating near its frequency
- C. It may cause atmospheric interference in the air around the antenna
- D. It may cause interference to other stations operating on a higher frequency band

B   Your equipment can cause spurious emissions if you operate it with some controls adjusted improperly. For example, if you operate an SSB transmitter with the microphone gain set too high you can overmodulate the signal, which will cause splatter interference on frequencies near the one on which you are operating.

T6C11 What may cause a buzzing or hum in the signal of an HF transmitter?

- A. Using an antenna that is the wrong length
- B. Energy from another transmitter
- C. Bad design of the transmitter's RF power output circuit
- D. A bad filter capacitor in the transmitter's power supply

D   As a licensed Amateur Radio operator, you are responsible for the quality of the signal transmitted from your station. The rules require your transmitted signal to be stable in frequency and pure in tone and modulation. A bad filter capacitor in your transmitter's power supply may result in ripple on the dc supply voltage. That in turn can cause a buzzing or hum on the signal from your HF transmitter.

T6C12 What is the major cause of telephone interference?

A. The telephone ringer is inadequate
B. Tropospheric ducting at UHF frequencies
C. The telephone was not equipped with interference protection when it was manufactured
D. Improper location of the telephone in the home

C   Radio frequency energy from your amateur transmitter may interfere with your own or your neighbor's consumer electronic equipment, and that includes telephones. As with receiver overload, there is nothing you can do at the transmitter to cure the interference. Interference protection measures must be taken at, or in, the device in question. The major cause of telephone interference comes from telephones that were not equipped with interference protection when they were manufactured. (Reference: FCC CIB Telephone Interference Bulletin)

Basic Communications Electronics

There will be 3 questions on your exam taken from the Basic Communications Electronics subelement printed in this chapter. These questions are divided into 3 groups, labeled T7A, T7B and T7C.

T7A **Fundamentals of electricity; AC/DC power; units and definitions of current, voltage, resistance, inductance, capacitance and impedance; Rectification; Ohm's Law principle (simple math); Decibel; Metric system and prefixes (e.g., pico, nano, micro, milli, deci, centi, kilo, mega, giga).**

T7A01 What is the name for the flow of electrons in an electric circuit?

   A. Voltage
   B. Resistance
   C. Capacitance
   D. Current

D   Electrons flow through the wires and components of an electric circuit. The flow of electrons in an electric circuit is called **current**.

T7A02 What is the name of a current that flows only in one direction?

   A. An alternating current
   B. A direct current
   C. A normal current
   D. A smooth current

B   **Direct current**, or **dc**, which flows only in one direction, travels from the negative battery terminal, through the circuit and back to the positive battery terminal.

T7A03 What is the name of a current that flows back and forth, first in one direction, then in the opposite direction?

    A. An alternating current
    B. A direct current
    C. A rough current
    D. A steady state current

A   **Alternating current,** or **ac**, alternates direction, flowing first in one direction, then in the opposite direction.

T7A04 What is the basic unit of electrical power?

    A. The ohm
    B. The watt
    C. The volt
    D. The ampere

B   The basic unit of electrical power is the **watt,** or **W**. This unit was named after James Watt (1736-1819), the inventor of the steam engine.

T7A05 What is the basic unit of electric current?

    A. The volt
    B. The watt
    C. The ampere
    D. The ohm

C   The basic unit of electric current, a measure of the rate of flow of electrons, is the **ampere,** abbreviated **A**. It is named for Andre Marie Ampere, an early 19th century scientist who studied electricity extensively.

T7A06 How much voltage does an automobile battery usually supply?

    A. About 12 volts
    B. About 30 volts
    C. About 120 volts
    D. About 240 volts

A   An automobile storage battery usually supplies about 12 **volts** of electrical potential.

T7A07 What limits the current that flows through a circuit for a particular applied DC voltage?

A. Reliance
B. Reactance
C. Saturation
D. Resistance

D   Resistance **opposes the flow of electrons** in an electrical circuit. We use **resistance** to control the amount of current that flows to various parts of a circuit given a particular applied dc voltage.

T7A08 What is the basic unit of resistance?

A. The volt
B. The watt
C. The ampere
D. The ohm

D   The **ohm** is the **basic unit** used to measure **circuit resistance**. It is named for Georg Simon Ohm, a German physicist and mathematician who discovered the relationship between voltage, current and resistance we call **Ohm's Law**.

T7A09 What is the basic unit of inductance?

A. The coulomb
B. The farad
C. The henry
D. The ohm

C   The **basic unit** of inductance is the **henry** (abbreviated H), named for the American physicist Joseph Henry. Because the henry is often too large for practical use in measurements, we use the **millihenry** (10^{-3}), abbreviated mH, or **microhenry** (10^{-6}), abbreviated μH.

T7A10 What is the basic unit of capacitance?

A. The farad
B. The ohm
C. The volt
D. The henry

A   The **basic unit** of capacitance is the **farad** (abbreviated F), named for Michael Faraday. Like the henry, the farad is too large a unit for practical measurements. For convenience, we use **microfarads** (10^{-6}), abbreviated μF, or **picofarads** (10^{-12}), abbreviated pF.

T7A11 Which of the following circuits changes an alternating current signal into a varying direct current signal?

A. Transformer
B. Rectifier
C. Amplifier
D. Director

B   A **rectifier** changes an alternating current signal into a varying direct current signal. Rectifiers are important components in radio receiver and transmitter circuitry.

T7A12 What formula shows how voltage, current and resistance relate to each other in an electric circuit?

A. Ohm's Law
B. Kirchhoff's Law
C. Ampere's Law
D. Tesla's Law

A   The relationship between circuit voltage, current and resistance is called **Ohm's Law. A circuit that has a current of 1 amp when a voltage of 1 volt is applied has a resistance of 1 ohm.**

T7A13 If a current of 2 amperes flows through a 50-ohm resistor, what is the voltage across the resistor?

- A. 25 volts
- B. 52 volts
- C. 100 volts
- D. 200 volts

C   This is an Ohm's Law question. Find the correct form of the Ohm's Law equation using the Ohm's Law Circle. Cover the letter representing the quantity you want to find and notice the positions of the other two letters. Since we want to find voltage in this question, you will see that the I and R are side by side, so you multiply them.

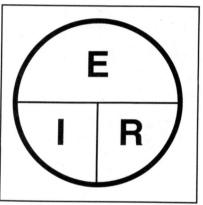

$$E = I \times R \quad E = 2\ A \times 50\ \Omega = 100\ V$$

On your calculator, enter this as:
2 × 50 = and read the answer as **100**.

Notice that we use the letter A to represent amperes, the unit for current. The letter V represents volts, the unit for voltage, and the Greek capital letter omega (Ω) represents ohms, the unit of resistance.

T7A14 If a 100-ohm resistor is connected to 200 volts, what is the current through the resistor?

- A. 1 ampere
- B. 2 amperes
- C. 300 amperes
- D. 20,000 amperes

B   See the explanation for question T7A13. This is an Ohm's Law question. Find the correct form of the Ohm's Law equation using the Ohm's Law Circle. Cover the letter representing the quantity you want to find and notice the positions of the other two letters. Since we want to find current in this question, you will see that the E is above the R, so you divide them.

$$I = E / R \qquad I = 200\ V / 100\ \Omega = 2\ A$$

On your calculator, enter this as: **200 ÷ 100 =** and read the answer as **2**.

T7A15 If a current of 3 amperes flows through a resistor connected to 90 volts, what is the resistance?

    A. 3 ohms
    B. 30 ohms
    C. 93 ohms
    D. 270 ohms

B   See the explanation for question T7A13. This is an Ohm's Law question. Find the correct form of the Ohm's Law equation using the Ohm's Law Circle. Cover the letter representing the quantity you want to find and notice the positions of the other two letters. Since we want to find resistance in this question, you will see that the E is above the I, so you divide them.

$$R = E / I \qquad R = 90 \text{ V} / 3 \text{ A} = 30 \text{ } \Omega$$

On your calculator, enter this as: **90 ÷ 3 =** and read the answer as **30**.

T7A16 If you increase your transmitter output power from 5 watts to 10 watts, what decibel (dB) increase does that represent?

    A. 2 dB
    B. 3 dB
    C. 5 dB
    D. 10 dB

B   When **power is doubled**, for instance from 5 to 10 watts, the increase is **3 dB**. Similarly, if the power is **reduced by half**, the decrease is **−3dB**.

T7A17 If an ammeter marked in amperes is used to measure a 3000-milliampere current, what reading would it show?

    A. 0.003 amperes
    B. 0.3 amperes
    C. 3 amperes
    D. 3,000,000 amperes

C   Use the chart to see how to change from milliamperes (mA) to amperes (A). The basic unit (U) on the chart represents amperes. Count three places to the left, and move the decimal point three places to the left to change 3000 mA to **3 A**.

10^9	10^6	10^3	10^2	10^1	10^0	10^{-1}	10^{-2}	10^{-3}	10^{-6}	10^{-9}	10^{-12}
G	• • M • • k		h	da	U	d	c	m • • μ • • n • • p			
giga	mega	kilo	hecto	deca	(unit)	deci	centi	milli	micro	nano	pico

T7A18 How many hertz are in a kilohertz?

    A. 10
    B. 100
    C. 1000
    D. 1,000,000

    C   Use the chart with question T7A17 to change from kilohertz (kHz) to hertz (H). The basic unit (U) on the chart represents hertz. Count three places to the right, and move the decimal point three places to the right to change 1 kHz to **1000 Hz**.

T7A19 If a dial marked in megahertz shows a reading of 3.525 MHz, what would it show if it were marked in kilohertz?

    A. 0.003525 kHz
    B. 35.25 kHz
    C. 3525 kHz
    D. 3,525,000 kHz

    C   Use the chart with question T7A17 to change from megahertz (MHz) to kilohertz (kHz). Find megahertz (10^6) on the chart. In converting the value 3.525 MHz to kHz, you count three steps to the right, and move the decimal point three places to the right, resulting in **3525 kHz**. Note that the two black dots between the mega and the kilo prefixes must be counted as these represent the two decimal places not shown between 10^6 and 10^3.

T7A20 How many microfarads is 1,000,000 picofarads?

    A. 0.001 microfarads
    B. 1 microfarad
    C. 1000 microfarads
    D. 1,000,000,000 microfarads

    B   Use the chart with question T7A17 to change from picofarads (pF) to microfarads (μF). Find the pico prefix at the far right of the chart. Count six places to the left (including the black dots) and move the decimal place 6 places to the left to change 1,000,000 pF to **1 μF**.

T7A21 If you have a hand-held transceiver with an output of 500 milliwatts, how many watts would this be?

    A. 0.02
    B. 0.5
    C. 5
    D. 50

B   Use the chart with question T7A17 to change from milliwatts to watts. To convert from 500 milliwatts to watts, find the milli prefix (10^{-3}) on the chart. Count three places to the left to the unit prefix, and move the decimal point three places to the left to change 500 mW to **0.5W**.

T7B   Basic electric circuits; Analog vs. digital communications; Audio/RF signal; Amplification.

T7B01 What type of electric circuit uses signals that can vary continuously over a certain range of voltage or current values?

    A. An analog circuit
    B. A digital circuit
    C. A continuous circuit
    D. A pulsed modulator circuit

A   An analog circuit carries signals that vary continuously over a range of current and voltage values. An example of an analog signal is the human voice as you hear it from the speaker of a communications receiver.

T7B02 What type of electric circuit uses signals that have voltage or current values only in specific steps over a certain range?

    A. An analog circuit
    B. A digital circuit
    C. A step modulator circuit
    D. None of these choices is correct

B   A **digital circuit** carries signals that have voltage and current values only in specific steps over a certain range. Examples of digital circuitry are the memory integrated circuits in your PC.

T7B03 Which of the following is an example of an analog communications method?

A. Morse code (CW)
B. Packet Radio
C. Frequency-modulated (FM) voice
D. PSK31

C Analog communications methods use continuously varying signal formats to transfer information. Examples of analog communications include AM and FM voice radio signal formats used in commercial broadcasting and the **frequency-modulated (FM) voice** signals used by radio amateurs.

T7B04 Which of the following is an example of a digital communications method?

A. Single-sideband (SSB) voice
B. Amateur Television (ATV)
C. FM voice
D. Radioteletype (RTTY)

D **Radioteletype (RTTY)** is an example of a digital communications method commonly used by radio amateurs. Another very common example is the Morse code transmissions heard on all of the HF amateur bands.

T7B05 Most humans can hear sounds in what frequency range?

A. 0 - 20 Hz
B. 20 - 20,000 Hz
C. 200 - 200,000 Hz
D. 10,000 - 30,000 Hz

B Most people can hear sounds in the frequency range of approximately **20 - 20,000 Hz**. Our ears respond to air-pressure changes, or vibrations in this frequency range.

T7B06 Why do we call electrical signals in the frequency range of 20 Hz to 20,000 Hz audio frequencies?

    A. Because the human ear cannot sense anything in this range
    B. Because the human ear can sense sounds in this range
    C. Because this range is too low for radio energy
    D. Because the human ear can sense radio waves in this range

B   The human ear generally hears sound in the range of approximately 20 to 20,000 Hz. Our ears respond to air-pressure changes, or vibations in this frequency range. Sounds in this range are **audible** to humans. This frequency range is called the **audio frequency** range. It is important to realize that our ears do not respond to electrical signals. If electrical signals in the range of 20 to 20,000 hertz are connected to a speaker, then the speaker will cause air-pressure variations that our ears can hear.

T7B07 What is the lowest frequency of electrical energy that is usually known as a radio frequency?

    A. 20 Hz
    B. 2,000 Hz
    C. 20,000 Hz
    D. 1,000,000 Hz

C   The radio frequency spectrum is generally considered to begin at **20,000 Hz (20 kHz)**. Although radio communications can take place on frequencies lower than 20 kHz, this is generally considered the lowest practical frequency for RF use.

T7B08 Electrical energy at a frequency of 7125 kHz is in what frequency range?

    A. Audio
    B. Radio
    C. Hyper
    D. Super-high

B   Electrical energy at a frequency of **7125 kHz** is considered in the **radio frequency range**. This frequency is allocated in the 40-meter amateur band in North America.

T7B09 If a radio wave makes 3,725,000 cycles in one second, what does this mean?

A. The radio wave's voltage is 3725 kilovolts
B. The radio wave's wavelength is 3725 kilometers
C. The radio wave's frequency is 3725 kilohertz
D. The radio wave's speed is 3725 kilometers per second

C   Cycles per second is the basic unit of frequency, also known as hertz (Hz). You can use the chart with question T7A17 to change from hertz (Hz) to kilohertz (kHz). The basic unit (U) on the chart represents hertz. Count three places to the left, and move the decimal point three places to the left to change **3,750,000 Hz** to **3725 kHz (kilohertz)**.

T7B10 Which component can amplify a small signal using low voltages?

A. A PNP transistor
B. A variable resistor
C. An electrolytic capacitor
D. A multiple-cell battery

A   A transistor is a solid-state device that can amplify a small signal using low voltages. Circuits using either NPN or **PNP transistors** can be designed for **small signal amplification** applications.

T7B11 Which component can amplify a small signal but normally uses high voltages?

A. A transistor
B. An electrolytic capacitor
C. A vacuum tube
D. A multiple-cell battery

C   Vacuum tube circuits can be built for **small signal amplification** applications. However, most vacuum tube circuits use **high voltages**.

T7C Concepts of Resistance/resistor; Capacitor/capacitance; Inductor/Inductance; Conductor/Insulator; Diode; Transistor; Semiconductor devices; Electrical functions of and schematic symbols of resistors, switches, fuses, batteries, inductors, capacitors, antennas, grounds and polarity; Construction of variable and fixed inductors and capacitors.

T7C01 Which of the following lists include three good electrical conductors?

A. Copper, gold, mica
B. Gold, silver, wood
C. Gold, silver, aluminum
D. Copper, aluminum, paper

C   In general, most **metals** make the best **conductors** of electricity. Gold, silver and aluminum are all excellent conductors, while wood, paper and mica are very poor conductors.

T7C02 What is one reason resistors are used in electronic circuits?

A. To block the flow of direct current while allowing alternating current to pass
B. To block the flow of alternating current while allowing direct current to pass
C. To increase the voltage of the circuit
D. To control the amount of current that flows for a particular applied voltage

D   A **resistor** controls the amount of current that flows in a circuit for a particular applied voltage as defined by **Ohm's Law**.

T7C03 If two resistors are connected in series, what is their total resistance?

A. The difference between the individual resistor values
B. Always less than the value of either resistor
C. The product of the individual resistor values
D. The sum of the individual resistor values

D   The **total resistance** of any number of resistors connected in **series** is the **sum** of the resistance values of each individual resistor.

T7C04 What is one reason capacitors are used in electronic circuits?

    A. To block the flow of direct current while allowing alternating current to pass

    B. To block the flow of alternating current while allowing direct current to pass

    C. To change the time constant of the applied voltage

    D. To change alternating current to direct current

A   One of the major functions of a capacitor in electronic circuits is to **allow alternating current to pass** while **blocking the passage of direct current**.

T7C05 If two equal-value capacitors are connected in parallel, what is their total capacitance?

    A. Twice the value of one capacitor

    B. Half the value of one capacitor

    C. The same as the value of either capacitor

    D. The value of one capacitor times the value of the other

A   The **total capacitance** of any number of capacitors connected in **parallel** is the **sum** of the capacitance values of each individual capacitor. Therefore, if two equal value capacitors were connected in parallel, their total capacitance would be twice the value of one capacitor.

T7C06 What does a capacitor do?

    A. It stores energy electrochemically and opposes a change in current

    B. It stores energy electrostatically and opposes a change in voltage

    C. It stores energy electromagnetically and opposes a change in current

    D. It stores energy electromechanically and opposes a change in voltage

B   A capacitor **stores energy** as an **electrostatic field** between the capacitor plates. This process **opposes** a change in the **voltage** across the capacitor.

T7C07 Which of the following best describes a variable capacitor?

    A. A set of fixed capacitors whose connections can be varied

    B. Two sets of insulating plates separated by a conductor, which can be varied in distance from each other

    C. A set of capacitors connected in a series-parallel circuit

    D. Two sets of rotating conducting plates separated by an insulator, which can be varied in surface area exposed to each other

D   The key element of a variable capacitor is the ability to vary the amount of conductor plate **surface area** exposed to each other. Since the capacitance value of any capacitor is related to the amount of conductor surface area on each plate, as the **plate spacing** is varied, so is the value of the capacitance of the device.

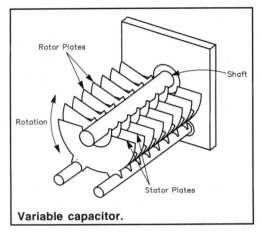

Rotor Plates

Shaft

Rotation

Stator Plates

Variable capacitor.

T7C08 What does an inductor do?

    A. It stores energy electrostatically and opposes a change in voltage

    B. It stores energy electrochemically and opposes a change in current

    C. It stores energy electromagnetically and opposes a change in current

    D. It stores energy electromechanically and opposes a change in voltage

C   An inductor **stores energy** as an **electromagnetic field**. This process **opposes** a change in the **current** through the inductor.

T7C09 What component controls current to flow in one direction only?

    A. A fixed resistor

    B. A signal generator

    C. A diode

    D. A fuse

C   A **diode** is a **semiconductor** device that conducts current in only one direction.

T7C10 What is one advantage of using ICs (integrated circuits) instead of vacuum tubes in a circuit?

A. ICs usually combine several functions into one package
B. ICs can handle high-power input signals
C. ICs can handle much higher voltages
D. ICs can handle much higher temperatures

A   IC's are made with semiconductor materials. An IC usually peforms **several circuit functions** in one package, and sometimes an entire application requires only the addition of a few external components. In addition, they use low operating voltages and are generally low-power devices.

T7C11 Which symbol of Figure T7-1 represents a fixed resistor?

A. Symbol 1
B. Symbol 2
C. Symbol 3
D. Symbol 5

C   The **fixed resistor** in this schematic diagram is represented by **symbol 3**. Fixed resistor symbols do not contain an arrow, as does a variable resistor. Note also that another fixed resistor appears in the diagram as **symbol 9**.

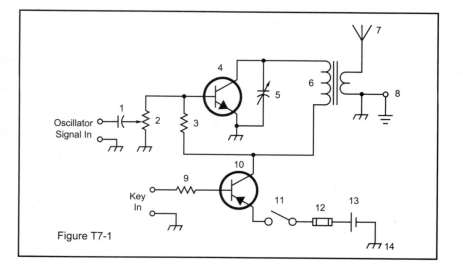

Figure T7-1

T7C12 In Figure T7-1, which symbol represents a variable resistor or potentiometer?

    A. Symbol 1
    B. Symbol 2
    C. Symbol 3
    D. Symbol 12

B   See the explanation for question T7C11. The **variable resistor**, also known as a potentiometer, in this schematic diagram is represented by **symbol 2**. A variable resistor will always contain a third terminal with an arrow pointed at the resistor body.

T7C13 In Figure T7-1, which symbol represents a single-cell battery?

    A. Symbol 1
    B. Symbol 6
    C. Symbol 12
    D. Symbol 13

D   See the explanation for question T7C11. The **battery** in this schematic is represented by **symbol 13**. A single-cell battery is generally shown with one long and one short vertical bar. Multiple cell batteries are shown with more than one pair of long and short vertical bars.

T7C14 In Figure T7-1, which symbol represents an NPN transistor?

    A. Symbol 2
    B. Symbol 4
    C. Symbol 10
    D. Symbol 12

B   See the explanation for question T7C11. The **NPN transistor** is represented by **symbol 4**. NPN transistors are always shown with the **arrow pointed away** from the junction. The arrow represents the direction of current flow through the base-emitter junction and out the emitter terminal.

T7C15 Which symbol of Figure T7-1 represents a fixed-value capacitor?

   A. Symbol 1
   B. Symbol 3
   C. Symbol 5
   D. Symbol 13

A   See the explanation for question T7C11. The **fixed value capacitor** is represented by **symbol 1**. Fixed valued capacitors are always represented with one straight and one curved line. There are no arrows in fixed valued capacitor symbols.

T7C16 In Figure T7-1, which symbol represents an antenna?

   A. Symbol 5
   B. Symbol 7
   C. Symbol 8
   D. Symbol 14

B   See the explanation for question T7C11. The **antenna** in the schematic drawing is represented by **symbol 7**. This symbol is intended to represent a generic antenna, rather than a specific kind of antenna.

T7C17 In Figure T7-1, which symbol represents a fixed-value iron-core inductor?

   A. Symbol 6
   B. Symbol 9
   C. Symbol 11
   D. Symbol 12

A   See the explanation for question T7C11. **Symbol 6** in the schematic drawing represents a **fixed-value iron-core inductor**. Note that this symbol also contains a second coil winding which connects to the antenna and ground terminals.

T7C18 In Figure T7-2, which symbol represents a single-pole, double-throw switch?

    A. Symbol 1
    B. Symbol 2
    C. Symbol 3
    D. Symbol 4

  D   The **single-pole, double-throw switch** is represented by **symbol 4** in Figure T7-2. A single-pole switch always has only one set of contacts, while a double-pole switch has two sets of contacts.

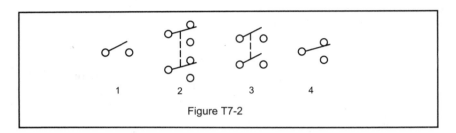

Figure T7-2

T7C19 In Figure T7-2, which symbol represents a double-pole, single-throw switch?

    A. Symbol 1
    B. Symbol 2
    C. Symbol 3
    D. Symbol 4

  C   The **double-pole, single-throw** switch is represented by **symbol 3** in Figure T7-2. The single-throw switch is distinguished from the double-throw switch by having only one output terminal, rather than two. This is basically an 'off-on' switch for one circuit only, while the double-throw switch can switch a common connection to one of two other possible circuits.

Good Engineering Practice

There will be 6 questions on your exam taken from the Good Engineering Practice subelement printed in this chapter. These questions are divided into 6 groups, labeled T8A through T8F.

T8A Basic amateur station apparatus; Choice of apparatus for desired communications; Setting up station; Constructing and modifying amateur station apparatus; Station layout for CW, SSB, FM, Packet and other popular modes.

T8A01 What two bands are most commonly used by "dual band" hand-held transceivers?

A. 6 meters and 2 meters
B. 2 meters and 1.25 meters
C. 2 meters and 70 cm
D. 70 cm and 23 cm

C   The two bands most commonly used by "dual band" hand-held transceivers are 2 meters and 70 cm. These bands are available to hams in Europe and Asia as well as in the Americas. The 1.25-meter band is not available to European and Asian hams, nor is the part of 6 meters where US hams operate FM.

T8A02 If your mobile transceiver works in your car but not in your home, what should you check first?

A. The power supply
B. The speaker
C. The microphone
D. The SWR meter

A   In your car, your mobile transceiver is usually powered directly from the car's battery. When you use the transceiver in your home, you'll need to connect it to a 12-V dc power supply. If your mobile transceiver works in your car but not in your home, you should check the power supply first.

T8A03 Which of the following devices would you need to conduct Amateur Radio communications using a data emission?

   A. A telegraph key
   B. A computer
   C. A transducer
   D. A telemetry sensor

B   Data communications are designed to be received and printed automatically. If you remember that, you should be able identify the computer as being the only device that would work with your radio to perform this function.

T8A04 Which of the following devices would be useful to create an effective Amateur Radio station for weak-signal VHF communication?

   A. A hand-held VHF FM transceiver
   B. A multi-mode VHF transceiver
   C. An Omni-directional antenna
   D. A mobile VHF FM transceiver

B   FM is not used for weak-signal VHF communications. SSB and CW are the modes that you would use. You'll need a multi-mode transceiver to operate those modes. For more effective communications with weak stations, you'll want to use a directional (gain) antenna.

T8A05 What would you connect to a transceiver for voice operation?

   A. A splatter filter
   B. A terminal-voice controller
   C. A receiver audio filter
   D. A microphone

D   For voice operation, you'll need a microphone to convert your voice to electrical impulses for transmission by your transceiver.

T8A06 What would you connect to a transceiver to send Morse code?

- A. A key-click filter
- B. A telegraph key
- C. An SWR meter
- D. An antenna switch

B To send Morse code with your transceiver, you'll need something to make the dots and dashes. Of the items listed, only the telegraph key will do that.

T8A07 What do many amateurs use to help form good Morse code characters?

- A. A key-operated on/off switch
- B. An electronic keyer
- C. A key-click filter
- D. A DTMF keypad

B An electronic keyer won't automatically send Morse code for you, but it will form perfect dots and dashes with perfect spacing between them. That's why many amateurs use an electronic keyer for sending Morse code.

T8A08 Why is it important to provide adequate power supply filtering for a CW transmitter?

- A. It isn't important, since CW transmitters cannot be modulated by AC hum
- B. To eliminate phase noise
- C. It isn't important, since most CW receivers can easily suppress any hum by using narrow filters
- D. To eliminate modulation of the RF signal by AC hum

D An inadequately filtered power supply will have ac hum in its output. That hum will modulate the RF signal output of a CW transmitter, and no amount of filtering at the receiver will remove the hum.

T8A09 Why is it important to provide adequate DC source supply filtering for a mobile transmitter or transceiver?

A. To reduce AC hum and carrier current device signals
B. To provide an emergency power source
C. To reduce stray noise and RF pick-up
D. To allow the use of smaller power conductors

C   Don't be confused by the previous question. In an automotive setup ac hum is not a problem like it can be when operating from your house. You may encounter stray noise and RF pick-up on the dc power lines to your mobile rig. Adequate filtering on these lines should eliminate the noise and RF.

T8A10 What would you connect to a transceiver for RTTY operation?

A. A modem and a teleprinter or computer system
B. A computer, a printer and a RTTY refresh unit
C. A data-inverter controller
D. A modem, a monitor and a DTMF keypad

A   To send and receive RTTY with your transceiver, you can use a mechanical teleprinter to generate the codes for transmission and to print received characters. Alternatively, a computer running appropriate software can handle the job. You'll also need a device to change codes to tones for transmission, and vice versa for reception. The drawing shows a typical RTTY station.

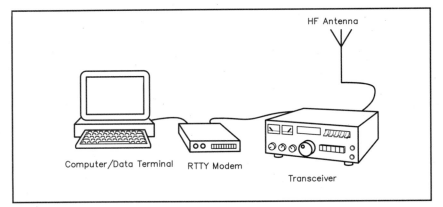

A typical RTTY station these days consists of a computer and a modem connected to a Transceiver.

T8A11 What might you connect between your transceiver and an antenna switch connected to several antennas?

A. A high-pass filter
B. An SWR meter
C. A key-click filter
D. A mixer

B   You might want to connect an SWR meter between your transceiver and an antenna switch. That will allow you to monitor your transmitter output and will give you an indication if you have switched to a wrong antenna. You'll want to place the SWR meter between the transceiver and the antenna switch so that you can check the performance of the antenna that you have selected.

T8A12 What might happen if you set your receiver's signal squelch too low while attempting to receive packet mode transmissions?

A. Noise may cause the TNC to falsely detect a data carrier
B. Weaker stations may not be received
C. Transmission speed and throughput will be reduced
D. The TNC could be damaged

A   If you set your receiver's signal squelch too low while operating packet radio, noise may cause the TNC to falsely detect a data carrier. This should not prevent reception, but it will prevent the TNC from switching into the transmit mode.

T8A13 What is one common method of transmitting RTTY on VHF/UHF bands?

A. Frequency shift the carrier to indicate mark and space at the receiver
B. Amplitude shift the carrier to indicate mark and space at the receiver
C. Key the transmitter on to indicate space and off for mark
D. Modulate a conventional FM transmitter with a modem

D   A common RTTY station for VHF/UHF uses a modem to modulate a conventional FM transmitter. Receive audio is also fed to the modem. A typical setup is illustrated in the drawing on the next page.

A typical 2-meter RTTY station may comprise an FM transceiver, a modem, and a computer. The modem is usually a multimode communications processor (MCP).

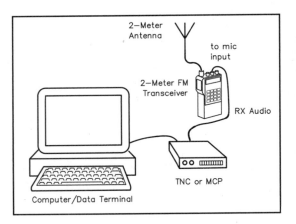

T8A14 What would you use to connect a dual-band antenna to a mobile transceiver that has separate VHF and UHF output connectors?

A. A dual-needle SWR meter
B. A full-duplex phone patch
C. Twin high-pass filters
D. A duplexer

D   A duplexer can be used to connect a dual-band antenna to a mobile transceiver that has separate VHF and UHF output connectors. See the drawing.

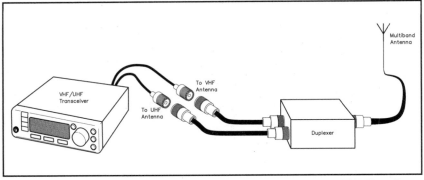

You can use a duplexer to connect a single antenna to a VHF/UHF radio that has two antenna connectors.

T8B How transmitters work; Operation and tuning; VFO; Transceiver; Dummy load; Antenna switch; Power supply; Amplifier; Stability; Microphone gain; FM deviation; Block diagrams of typical stations.

T8B01 Can a transceiver designed for FM phone operation also be used for single sideband in the weak-signal portion of the 2-meter band?

A. Yes, with simple modification
B. Only if the radio is a "multimode" radio
C. Only with the right antenna
D. Only with the right polarization

B   An FM transceiver will not operate SSB. It is not a simple matter to add SSB capability to an FM radio. However, many folks use a "multimode" radio to operate FM. A multimode radio has the necessary circuitry to operate both FM and SSB.

T8B02 How is a CW signal usually transmitted?

A. By frequency-shift keying an RF signal
B. By on/off keying an RF signal
C. By audio-frequency-shift keying an oscillator tone
D. By on/off keying an audio-frequency signal

B   A CW signal is usually transmitted by keying an RF carrier on and off. This is about the simplest way to generate the CW signal.

T8B03 What purpose does block 1 serve in the simple CW transmitter pictured in Figure T8-1?

A. It detects the CW signal
B. It controls the transmitter frequency
C. It controls the transmitter output power
D. It filters out spurious emissions from the transmitter

B   See Figure T8-1 on the next page. Block 1 is the oscillator of the simple CW transmitter. It can be a variable-frequency oscillator (VFO), or it may be a crystal controlled oscillator. In either case, the circuitry in block 1 controls the transmitter frequency.

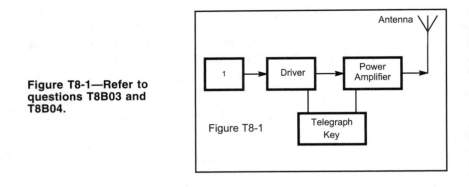

Figure T8-1—Refer to questions T8B03 and T8B04.

Figure T8-1

T8B04 **What circuit is pictured in Figure T8-1 if block 1 is a variable-frequency oscillator?**

    A. A packet-radio transmitter
    B. A crystal-controlled transmitter
    C. A single-sideband transmitter
    D. A VFO-controlled transmitter

    D   If block 1 is a variable-frequency oscillator, then the circuit is a VFO-controlled transmitter.

T8B05 **What circuit is shown in Figure T8-2 if block 1 represents a reactance modulator?**

    A. A single-sideband transmitter
    B. A double-sideband AM transmitter
    C. An FM transmitter
    D. A product transmitter

    C   See Figure T8-2 on the next page. If block 1 represents a reactance modulator, the circuit is that of an FM transmitter.

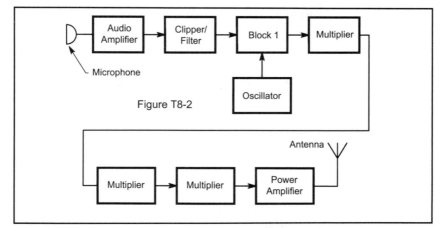

Figure T8-2—Refer to questions T8B05 and T8B06.

T8B06 How would the output of the FM transmitter shown in Figure T8-2 be affected if the audio amplifier failed to operate (assuming block 1 is a reactance modulator)?

A. There would be no output from the transmitter
B. The output would be 6-dB below the normal output power
C. The transmitted audio would be distorted but understandable
D. The output would be an unmodulated carrier

D The purpose of a reactance modulator is to shift the frequency from an oscillator according to the modulating audio. The result is an FM signal. If the audio amplifier failed to operate, there would be no audio going to the modulator. This would result in no change in the oscillator frequency. In other words, the transmitter output would be an unmodulated carrier.

T8B07 What minimum rating should a dummy antenna have for use with a 100-watt, single-sideband-phone transmitter?

A. 100 watts continuous
B. 141 watts continuous
C. 175 watts continuous
D. 200 watts continuous

A A dummy antenna for use with a transmitter should be capable of handling the full output of that transmitter. A dummy antenna with a minimum rating of 100 watts continuous would be suitable for use with a 100-watt, single-sideband-phone transmitter.

T8B08 A mobile radio may be operated at home with the addition of which piece of equipment?

    A. An alternator
    B. A power supply
    C. A linear amplifier
    D. A rhombic antenna

    B   A mobile radio is made to operate from the 12 V dc from your car's battery. You can use it in your house if you have a 12 V dc power supply to power the radio.

T8B09 What might you use instead of a power supply for home operation of a mobile radio?

    A. A filter capacitor
    B. An alternator
    C. A 12-volt battery
    D. A linear amplifier

    C   The purpose of the power supply is to provide 12 V dc to operate the mobile radio. Instead of the power supply, you could use a 12-volt battery.

T8B10 What device converts 120 V AC to 12 V DC?

    A. A catalytic converter
    B. A low-pass filter
    C. A power supply
    D. An RS-232 interface

    C   The device that takes the 120 V ac that you use in your house and changes it to 12 V dc is called a power supply. You would use this type of power supply to operate a mobile radio from your home.

T8B11 What device could boost the low-power output from your hand-held radio up to 100 watts?

    A. A voltage divider
    B. A power amplifier
    C. A impedance network
    D. A voltage regulator

    B   To boost the power output from your hand-held radio, you'll need a device to increase or amplify the output power. That device is called, logically enough, a power amplifier.

T8B12 What is the result of over deviation in an FM transmitter?

A. Increased transmitter power
B. Out-of-channel emissions
C. Increased transmitter range
D. Poor carrier suppression

B   Over deviation is caused by too much modulation. In other words, the carrier is being shifted more than it should and this means that the transmitted signal is wider than it should be. This will result in out-of-channel emissions.

T8B13 What can you do if you are told your FM hand-held or mobile transceiver is over deviating?

A. Talk louder into the microphone
B. Let the transceiver cool off
C. Change to a higher power level
D. Talk farther away from the microphone

D   If your radio is over deviating, it means that there is too much modulation. The question is how to reduce it. You will decrease the modulation level (deviation) if you lower your voice or talk farther away from the microphone.

T8B14 In Figure T8-3, if block 1 is a transceiver and block 3 is a dummy antenna, what is block 2?

A. A terminal-node switch
B. An antenna switch
C. A telegraph key switch
D. A high-pass filter

B   The question is looking for something that would go between a transceiver and an antenna and a dummy antenna. Block 2 is an antenna switch. Its function is to connect the transceiver output jack to either the antenna for regular operation or to the dummy antenna for testing.

**Figure T8-3—Refer to questions
T8B14 and T8B15.**

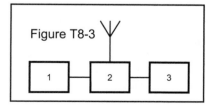

Figure T8-3

T8B15 In Figure T8-3, if block 1 is a transceiver and block 2 is an antenna switch, what is block 3?

- A. A terminal-node switch
- B. An SWR meter
- C. A telegraph key switch
- D. A dummy antenna

D   Based the previous question, you've probably deduced that block 3 is a dummy antenna. If so, you are 100% correct.

T8B16 In Figure T8-4, if block 1 is a transceiver and block 2 is an SWR meter, what is block 3?

- A. An antenna switch
- B. An antenna tuner
- C. A key-click filter
- D. A terminal-node controller

B   In the block diagram, block 3 is an antenna tuner. It is located between the SWR meter and the antenna. This allows you to see the SWR that the transceiver "sees" as you adjust the antenna tuner. Block 3 is where an antenna switch might go, but the switch would serve no useful function unless there were other antennas or dummy antennas.

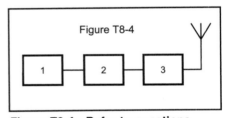

Figure T8-4

Figure T8-4—Refer to questions T8B16 through T8B18.

T8B17 In Figure T8-4, if block 1 is a transceiver and block 3 is an antenna tuner, what is block 2?

- A. A terminal-node switch
- B. A dipole antenna
- C. An SWR meter
- D. A high-pass filter

C   Use the same logic that you used for answering the previous question. You'll need an SWR meter between your transceiver and antenna tuner so that you can adjust the tuner properly.

T8B18 In Figure T8-4, if block 2 is an SWR meter and block 3 is an antenna tuner, what is block 1?

A. A terminal-node switch
B. A power supply
C. A telegraph key switch
D. A transceiver

D   The correct order is transceiver, SWR meter, antenna tuner and antenna.

T8C   How receivers work, operation and tuning, including block diagrams; Superheterodyne including Intermediate frequency; Reception; Demodulation or Detection; Sensitivity; Selectivity; Frequency standards; Squelch and audio gain (volume) control.

T8C01 What type of circuit does Figure T8-5 represent if block 1 is a product detector?

A. A simple phase modulation receiver
B. A simple FM receiver
C. A simple CW and SSB receiver
D. A double-conversion multiplier

C   A product detector is used for receiving SSB or CW signals. It will not work for FM or PM.

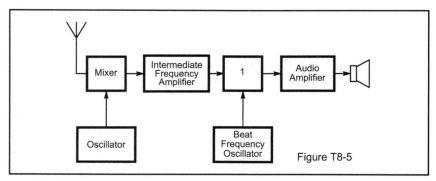

Figure T8-5—Refer to questions T8C01 and T8C02.

T8C02 If Figure T8-5 is a diagram of a simple single-sideband receiver, what type of circuit should be shown in block 1?

A. A high pass filter
B. A ratio detector
C. A low pass filter
D. A product detector

D   See the drawing with question T8C01. The SSB receiver needs a product detector at block 1. The circuit in block 1 mixes a signal from the IF amplifier with the output of the BFO and gives you an audio output. That's what a product detector does.

T8C03 What circuit is pictured in Figure T8-6, if block 1 is a frequency discriminator?

A. A double-conversion receiver
B. A variable-frequency oscillator
C. A superheterodyne receiver
D. An FM receiver

D   A frequency discriminator is used as a detector for FM or phase-modulated signals. That fact should help you identify the block diagram as an FM receiver.

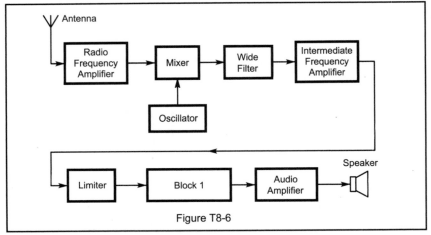

Figure T8-6

Figure T8-6—Refer to questions T8C03 through T8C05.

T8C04 What is block 1 in the FM receiver shown in Figure T8-6?

    A. A frequency discriminator
    B. A product detector
    C. A frequency-shift modulator
    D. A phase inverter

A   See the drawing with question T8C03. The FM receiver needs a detector, and that's what goes in block 1. For FM signals, you'll need a frequency discriminator to act as your detector. The other circuits will not do that job.

T8C05 What would happen if block 1 failed to function in the FM receiver diagram shown in Figure T8-6?

    A. The audio output would sound loud and distorted
    B. There would be no audio output
    C. There would be no effect
    D. The receiver's power supply would be short-circuited

B   See the drawing with question T8C03. Block 1 is the frequency discriminator, which acts as the receiver's detector stage. The function of that stage is to extract audio from the modulated RF signal that you are receiving. If the circuit in block 1 fails, audio is not extracted and that means that there would be no audio output.

T8C06 What circuit function is found in all types of receivers?

    A. An audio filter
    B. A beat-frequency oscillator
    C. A detector
    D. An RF amplifier

C   You might find any of these functions in a particular receiver. However, it is only the detector stage that can be found in all receivers. The job of a detector is to extract audio or data from a modulated signal.

T8C07 What is one accurate way to check the calibration of your receiver's tuning dial?

A. Monitor the BFO frequency of a second receiver
B. Tune to a popular amateur net frequency
C. Tune to one of the frequencies of station WWV or WWVH
D. Tune to another amateur station and ask what frequency the operator is using

C Radio stations WWV and WWVH are operated by the US government. Their frequencies are precisely controlled, and that makes them ideal checkpoints when calibrating your receiver's tuning dial.

T8C08 What circuit combines signals from an IF amplifier stage and a beat-frequency oscillator (BFO), to produce an audio signal?

A. An AGC circuit
B. A detector circuit
C. A power supply circuit
D. A VFO circuit

B The job of a detector is to extract audio or data from a modulated signal. Of the circuits listed, only the detector circuit produces an audio output.

T8C09 Why is FM voice so effective for local VHF/UHF radio communications?

A. The carrier is not detectable
B. It is more resistant to distortion caused by reflected signals than the AM modes
C. It has audio that is less affected by interference from static-type electrical noise than the AM modes
D. Its RF carrier stays on frequency better than the AM modes

C Noise impulses are characterized by changes in their amplitude. There is little change in the frequency components of that type noise. When operating AM modes, such as SSB, you will detect that noise in your receiver. When signals are strong, as in local communications, FM voice is nearly noise free. In other words, FM has audio that is less affected by interference from static-type electrical noise than the AM modes.

T8C10 Why do many radio receivers have several IF filters of different bandwidths that can be selected by the operator?

A. Because some frequency bands are wider than others
B. Because different bandwidths help increase the receiver sensitivity
C. Because different bandwidths improve S-meter readings
D. Because some emission types need a wider bandwidth than others to be received properly

D   Wider filters allow more noise and interference to pass through to the detector. For that reason you'll normally want to use an IF filter that's just wide enough for the type of signal that your are receiving. A bandwidth of less than 100 Hz will work fine for CW, but for data modes you'll probably need more bandwidth. For SSB signals, you need only half the bandwidth that you do for double sideband with carrier. It all boils down to this, some emission types need a wider bandwidth than others to be received properly.

T8C11 What is the function of a mixer in a superheterodyne receiver?

A. To cause all signals outside of a receiver's passband to interfere with one another
B. To cause all signals inside of a receiver's passband to reinforce one another
C. To shift the frequency of the received signal so that it can be processed by IF stages
D. To interface the receiver with an auxiliary device, such as a TNC

C   The function of a mixer is to shift the frequency of the received signal to the intermediate frequency (IF). After being shifted, the received signal can be processed by the receiver's IF stages.

T8C12 What frequency or frequencies could the radio shown in Figure T8-7 receive?

A. 136.3 MHz
B. 157.7 MHz and 10.7 MHz
C. 10.7 MHz
D. 147.0 MHz and 168.4 MHz

D   When two signals are fed into a mixer, two new signals appear at the mixer output. These new signals are at the sum of the two frequencies and at the difference of the two. In this case, the receive frequency of the radio will be 157.7 MHz plus 10.7 MHz or 157.7 MHz minus 10.7 MHz. That means that 147.0 MHz or 168.4 MHz could be received by the radio shown in the drawing on the next page.

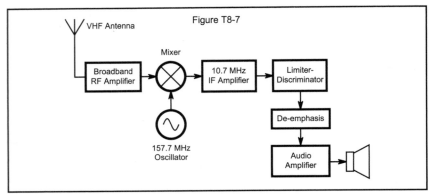

Figure T8-7

VHF Antenna

Mixer

Broadband RF Amplifier

10.7 MHz IF Amplifier

Limiter-Discriminator

De-emphasis

157.7 MHz Oscillator

Audio Amplifier

Figure T8-7—Refer to questions T8C12 through T8C14.

T8C13 What type of receiver is shown in Figure T8-7?

A. Direct conversion
B. Superregenerative
C. Single-conversion superheterodyne
D. Dual conversion superheterodyne

C   The receiver has an RF amplifier and an oscillator feeding into a mixer. From the mixer signals go through an IF amplifier and on to the detector. This is the classic lineup for a single-conversion superheterodyne receiver. It's a superheterodyne because the receive frequency is shifted to an intermediate frequency. It's single conversion because it only does this shifting once as the signal passes through the receiver.

T8C14 What emission mode could the receiver in Figure T8-7 detect?

A. AM
B. FM
C. Single sideband (SSB)
D. CW

B   The receiver's detector determines the receiver's emission mode. The receiver in the figure has a limiter and frequency discriminator for a detector. That means it is an FM receiver.

T8C15 Where should the squelch be set for the proper operation of an FM receiver?

- A. Low enough to hear constant background noise
- B. Low enough to hear chattering background noise
- C. At the point that just silences background noise
- D. As far beyond the point of silence as the knob will turn

C   The nice thing about FM voice operation is that it is virtually noise free while receiving a strong signal. The same cannot be said when there is no signal present. In the absence of a signal you will hear noise. Most FM receivers have a squelch control that cuts off this noise when the incoming signal is below a level set by the squelch control. The proper squelch setting is enough to silence the background noise and no more. If you go beyond that point, you may not be able to hear weaker signals. Far enough beyond and you may not hear strong signals.

T8D   **How antennas work; Radiation principles; Basic construction; Half wave dipole length vs. frequency; Polarization; Directivity; ERP; Directional/non-directional antennas; Multiband antennas; Antenna gain; Resonant frequency; Loading coil; Electrical vs. physical length; Radiation pattern; Transmatch.**

T8D01 Which of the following will improve the operation of a hand-held radio inside a vehicle?

- A. Shielding around the battery pack
- B. A good ground to the belt clip
- C. An external antenna on the roof
- D. An audio amplifier

C   When your hand-held radio is inside a vehicle, so is its antenna. That won't totally shield the antenna, but it will reduce its performance. You can improve you communications effectiveness by using an external antenna, and that will work best if it's mounted on the roof of your vehicle.

T8D02 Which is true of "rubber duck" antennas for hand-held transceivers?

    A. The shorter they are, the better they perform

    B. They are much less efficient than a quarter-wavelength telescopic antenna

    C. They offer the highest amount of gain possible for any hand-held transceiver antenna

    D. They have a good long-distance communications range

B   When you buy a new VHF hand-held transceiver, it will usually have a flexible rubber antenna commonly called a "rubber duck." This antenna is inexpensive, small, lightweight and difficult to break. On the other hand, it has some disadvantages: it is a compromise design that is inefficient and thus does not perform as well as larger antenna, such as a quarter-wavelength telescopic whip antenna.

T8D03 What would be the length, to the nearest inch, of a half-wavelength dipole antenna that is resonant at 147 MHz?

    A. 19 inches

    B. 37 inches

    C. 55 inches

    D. 74 inches

B   You'll need to know the length formula for a half-wavelength dipole, which is

$$\text{Length (in feet)} = \frac{468}{f\,(\text{MHz})}$$

Calculate the length in inches by multiplying the length in feet by 12. Now, plug the numbers into the equation and you get

$$\text{Length (inches)} = \frac{468}{147} \times 12 = 38.2 \text{ inches}$$

The best choice among those given is 37 inches, which makes it the right answer.

T8D04 How long should you make a half-wavelength dipole antenna for 223 MHz (measured to the nearest inch)?

    A. 112 inches
    B. 50 inches
    C. 25 inches
    D. 12 inches

C   Use the same formula to calculate the length of a half-wavelength dipole.

$$\text{Length (inches)} = \frac{468}{223} \times 12 = 25.2 \text{ inches}$$

Round the results to the nearest inch and you get 25 inches.

T8D05 How long should you make a quarter-wavelength vertical antenna for 146 MHz (measured to the nearest inch)?

    A. 112 inches
    B. 50 inches
    C. 19 inches
    D. 12 inches

C   This time you need to calculate the length of a quarter-wavelength vertical antenna. That means it's half the length of a half-wavelength dipole. Calculate the length with the formula after you substitute 234 for 468. Then round off the results to the nearest inch.

$$\text{Length (inches)} = \frac{234}{146} \times 12 = 19 \text{ inches}$$

T8D06 How long should you make a quarter-wavelength vertical antenna for 440 MHz (measured to the nearest inch)?

    A. 12 inches
    B. 9 inches
    C. 6 inches
    D. 3 inches

C   Use the same formula that you did for the previous question.

$$\text{Length (inches)} = \frac{234}{440} \times 12 = 6 \text{ inches}$$

T8D07 **Which of the following factors has the greatest effect on the gain of a properly designed Yagi antenna?**

A. The number of elements
B. Boom length
C. Element spacing
D. Element diameter

B Although several factors affect the amount of gain of a Yagi antenna, boom length has the greatest effect: the longer the boom, the higher the gain.

T8D08 **Approximately how long is the driven element of a Yagi antenna?**

A. 1/4 wavelength
B. 1/3 wavelength
C. 1/2 wavelength
D. 1 wavelength

C The driven element of a Yagi antenna is about ½ wavelength long at the antenna design frequency. The feed line connects to this driven element.

T8D09 **In Figure T8-8, what is the name of element 2 of the Yagi antenna?**

A. Director
B. Reflector
C. Boom
D. Driven element

D Element 2 of the Yagi antenna in Figure T8-8 is the driven element. It gets its name from the fact that the feed line connects to (drives) that element.

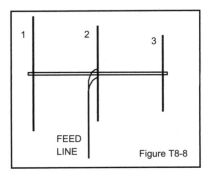

Figure T8-8—Refer to questions T8D09 through T8D11.

T8D10 In Figure T8-8, what is the name of element 3 of the Yagi antenna?

    A. Director
    B. Reflector
    C. Boom
    D. Driven element

A   Element 3 of the Yagi antenna in Figure T8-8 is called a director. Directors are a bit shorter than the driven element. The direction of maximum radiation from a Yagi antenna is from the reflector through the driven element to the director.

T8D11 In Figure T8-8, what is the name of element 1 of the Yagi antenna?

    A. Director
    B. Reflector
    C. Boom
    D. Driven element

B   Element 1 of the Yagi antenna in Figure T8-8 is called the reflector. It is a bit longer than the driven element.

T8D12 What is a cubical quad antenna?

    A. Four straight, parallel elements in line with each other, each approximately 1/2-electrical wavelength long
    B. Two or more parallel four-sided wire loops, each approximately one-electrical wavelength long
    C. A vertical conductor 1/4-electrical wavelength high, fed at the bottom
    D. A center-fed wire 1/2-electrical wavelength long

B   The cubical quad antenna is a type of beam antenna that uses parasitic elements. The elements of a quad antenna are usually wire loops. The total length of the wire in the elements is approximately one electrical wavelength. The elements of the quad are usually square, which means that each side of the square is about ¼-wavelength long.

T8D13 What does horizontal wave polarization mean?

A. The magnetic lines of force of a radio wave are parallel to the Earth's surface
B. The electric lines of force of a radio wave are parallel to the Earth's surface
C. The electric lines of force of a radio wave are perpendicular to the Earth's surface
D. The electric and magnetic lines of force of a radio wave are perpendicular to the Earth's surface

B   Polarization refers to the electrical-field characteristic of a radio wave. You can think of it as how the antenna is positioned. An antenna that is parallel to the Earth's surface (like a dipole antenna) produces horizontally polarized radio waves.

T8D14 What does vertical wave polarization mean?

A. The electric lines of force of a radio wave are parallel to the Earth's surface
B. The magnetic lines of force of a radio wave are perpendicular to the Earth's surface
C. The electric lines of force of a radio wave are perpendicular to the Earth's surface
D. The electric and magnetic lines of force of a radio wave are parallel to the Earth's surface

C   Vertical wave polarization means that the electric lines of force of a radio wave are perpendicular (at a 90° angle) to the Earth's surface. An antenna that is perpendicular to the earth's surface, such as a ¼-wavelength vertical, produces vertically polarized waves.

T8D15 If the ends of a half-wavelength dipole antenna (mounted at least a half-wavelength high) point east and west, which way would the antenna send out radio energy?

A. Equally in all directions
B. Mostly up and down
C. Mostly north and south
D. Mostly east and west

C   Dipole antennas send radio energy best in a direction that is 90° to the antenna wire. For example, suppose you install a dipole antenna so the ends of the wire run in an east/west direction. Assuming it was well off the ground (preferably ½ wavelength high), this antenna would send stronger signals in north and south directions.

T8D16 What electromagnetic wave polarization do most repeater antennas have in the VHF and UHF spectrum?

    A. Horizontal
    B. Vertical
    C. Right-hand circular
    D. Left-hand circular

B   Most VHF and UHF repeater antennas use vertical polarization. Simple omnidirectional vertical whip antennas work well for working into these repeaters.

T8D17 What electromagnetic wave polarization is used for most satellite operation?

    A. Only horizontal
    B. Only vertical
    C. Circular
    D. No polarization

C   In addition to horizontal and vertical, there is one more type of polarization: circular. Signals from orbiting satellites are circularly polarized, so the ground-based antennas that receive these signals should also be circularly polarized.

T8D18 Which antenna polarization is used most often for weak signal VHF/UHF SSB operation?

    A. Vertical
    B. Horizontal
    C. Right-hand circular
    D. Left-hand circular

B   For weak signal VHF/UHF SSB operation, most operators use antennas with horizontal polarization. This usually means horizontal beam antennas such as Yagis.

T8D19 How will increasing antenna gain by 3 dB affect your signal's effective radiated power in the direction of maximum radiation?

A. It will cut it in half
B. It will not change
C. It will double it
D. It will quadruple it

C   A two-to-one power ratio represents a 3 dB difference. Increasing your antenna gain by 3 dB has the same effect as doubling your transmit power. A 3-dB increase in your antenna gain doubles your signal's effective radiated power.

T8D20 What is one advantage to using a multiband antenna?

A. You can operate on several bands with a single feed line
B. Multiband antennas always have high gain
C. You can transmit on several frequencies simultaneously
D. Multiband antennas offer poor harmonic suppression

A   The principal advantage to using a multiband antenna is that it will operate on several bands with a single feed line.

T8D21 What could be done to reduce the physical length of an antenna without changing its resonant frequency?

A. Attach a balun at the feed point
B. Add series capacitance at the feed point
C. Use thinner conductors
D. Add a loading coil

D   Shorter antennas have higher resonant frequencies, but you can lower the resonant frequency by inserting a loading coil into the antenna. In the case of verticals, loading coils are typically placed either at the base of the vertical for easy access and pruning, or about halfway up the antenna, where the efficiency of the antenna is improved but access for fine-tuning may be more difficult.

T8D22 What device might allow use of an antenna on a band it was not designed for?

A. An SWR meter
B. A low-pass filter
C. An antenna tuner
D. A high-pass filter

C   A useful accessory that you will see in many ham shacks is an antenna tuner. This device may let you use one antenna on several bands. The matching network may also allow you to use your antenna on a band it is not designed for. The antenna tuner matches (tunes) the impedance of the load (the antenna and feed line) to the impedance of your transmitter.

T8E   How transmission lines work; Standing waves/SWR/ SWR-meter; Impedance matching; Types of transmission lines; Feed point; Coaxial cable; Balun; Waterproofing Connections.

T8E01 What does standing-wave ratio mean?

A. The ratio of maximum to minimum inductances on a feed line
B. The ratio of maximum to minimum capacitances on a feed line
C. The ratio of maximum to minimum impedances on a feed line
D. The ratio of maximum to minimum voltages on a feed line

D   Standing wave ratio, or SWR, is defined as the ratio of the maximum voltage to the minimum voltage along a transmission line. Voltage differences are an indication of reflected power, which is caused by an impedance mismatch between the line and load. An SWR of 1:1 means you have no reflected power. The transmission line in that case is said to be "flat."

T8E02 What instrument is used to measure standing wave ratio?

A. An ohmmeter
B. An ammeter
C. An SWR meter
D. A current bridge

C   Use an SWR meter to measure the standing wave ratio. Most hams connect an SWR meter between the transmitter output and an antenna switch or antenna tuner. This gives an indication of the impedance match between the antenna system—including feed line and antenna—and the transmitter.

T8E03 What would an SWR of 1:1 indicate about an antenna system?

    A. That the antenna was very effective
    B. That the transmission line was radiating
    C. That the antenna was reflecting as much power as it was radiating
    D. That the impedance of the antenna and its transmission line were matched

D   An SWR of 1:1 indicates that the impedance of the antenna and its transmission line are perfectly matched. That means no standing waves. The minimum voltage and the maximum voltage are the same. The ratio of the two is 1, and that's as good as it gets.

T8E04 What does an SWR reading of 4:1 mean?

    A. An impedance match that is too low
    B. An impedance match that is good, but not the best
    C. An antenna gain of 4
    D. An impedance mismatch; something may be wrong with the antenna system

D   An SWR reading of 4:1 indicates a significant impedance mismatch. There may be something wrong with your antenna system. Normally, you'll want to have an SWR of 2:1 or better.

T8E05 What does an antenna tuner do?

    A. It matches a transceiver output impedance to the antenna system impedance
    B. It helps a receiver automatically tune in stations that are far away
    C. It switches an antenna system to a transceiver when sending, and to a receiver when listening
    D. It switches a transceiver between different kinds of antennas connected to one feed line

A   Your transmitter may not operate very well if it is connected to a feed line that presents an impedance that is significantly different from the transmitter output impedance. Your transmitter probably has an output circuit designed for a 50-ohm load. Your antenna system (the combination of your antenna and feed line) may not have an impedance of 50 ohms. An antenna tuner matches your transceiver's output impedance to the antenna system impedance.

T8E06 What is a coaxial cable?

A. Two wires side-by-side in a plastic ribbon
B. Two wires side-by-side held apart by insulating rods
C. Two wires twisted around each other in a spiral
D. A center wire inside an insulating material covered by a metal sleeve or shield

D The most common type of feed line used by amateurs is called "coax" (pronounced kó-aks) for short. This feed line has one conductor inside the other. It's like a wire inside a flexible tube. The center conductor is surrounded by insulation, and the insulation is surrounded by a wire braid called the shield. The whole cable is then encased in a tough vinyl outer coating, which makes the cable weatherproof.

T8E07 Why should you use only good quality coaxial cable and connectors for a UHF antenna system?

A. To keep RF loss low
B. To keep television interference high
C. To keep the power going to your antenna system from getting too high
D. To keep the standing-wave ratio of your antenna system high

A Any line that feeds an antenna absorbs a small amount of transmitter power. That power is lost, because it serves no useful purpose. (The lost power warms the feed line slightly.) Feed line losses increase as frequency increases. Better-quality coaxial cables have lower loss than poor-quality cables. More of the transmitter power is lost as heat in a poor-quality coaxial cable.

T8E08 What is parallel-conductor feed line?

A. Two wires twisted around each other in a spiral
B. Two wires side-by-side held apart by insulating material
C. A center wire inside an insulating material that is covered by a metal sleeve or shield
D. A metal pipe that is as wide or slightly wider than a wavelength of the signal it carries

B In open-wire line, also known as parallel-conductor feed line, two parallel wires are spaced a constant distance from each other by insulation of some kind. One of the main advantages of open-wire transmission line is that the loss is less than that for coaxial cable. Open-wire line is also considerably cheaper than coax. But it has the disadvantage that it must be kept away from metallic objects.

T8E09 Which of the following are some reasons to use parallel-conductor, open-wire feed line?

A. It has low impedance and will operate with a high SWR
B. It will operate well even with a high SWR and it works well when tied down to metal objects
C. It has a low impedance and has less loss than coaxial cable
D. It will operate well even with a high SWR and has less loss than coaxial cable

D   Open-wire feed line has less loss than coaxial cable. It has the further advantage that it will operate well even with a high SWR.

T8E10 What does "balun" mean?

A. Balanced antenna network
B. Balanced unloader
C. Balanced unmodulator
D. Balanced to unbalanced

D   When connecting coaxial (unbalanced) feed lines to balanced antennas, many hams use a balun. Balun is a contraction for *bal*anced to *un*balanced.

T8E11 Where would you install a balun to feed a dipole antenna with 50-ohm coaxial cable?

A. Between the coaxial cable and the antenna
B. Between the transmitter and the coaxial cable
C. Between the antenna and the ground
D. Between the coaxial cable and the ground

A   When connecting coaxial (unbalanced) feed lines to balanced antennas, many hams use a balun. You install the balun at the antenna feed point, between the coaxial cable and the antenna.

T8E12 What happens to radio energy when it is sent through a poor quality coaxial cable?

A. It causes spurious emissions
B. It is returned to the transmitter's chassis ground
C. It is converted to heat in the cable
D. It causes interference to other stations near the transmitting frequency

C   Any line that feeds an antenna absorbs a small amount of transmitter power. That power is lost and only serves to warm the feed line slightly. More of the transmitter power is lost as heat in a poor-quality coaxial cable.

T8E13 What is an unbalanced line?

A. A feed line with neither conductor connected to ground
B. A feed line with both conductors connected to ground
C. A feed line with one conductor connected to ground
D. All of these answers are correct

C   An unbalanced line is a feed line with one conductor connected to ground. Coax cable is an unbalanced line that is commonly used by amateurs.

T8E14 What point in an antenna system is called the feed point?

A. The antenna connection on the back of the transmitter
B. Halfway between the transmitter and the feed line
C. At the point where the feed line joins the antenna
D. At the tip of the antenna

C   The feed line connects to an antenna at its feed point. The feed point may be in the center of the antenna, at one end or anywhere in between.

## T8F	Voltmeter/ammeter/ohmmeter/multi/S-meter, peak reading and RF wattmeter; Building/modifying equipment; Soldering; Making measurements; Test instruments.

T8F01 Which instrument would you use to measure electric potential or electromotive force?

- A. An ammeter
- B. A voltmeter
- C. A wavemeter
- D. An ohmmeter

B   The unit of electric potential (or electromotive force) is the volt. You would use a voltmeter to measure electric potential.

T8F02 How is a voltmeter usually connected to a circuit under test?

- A. In series with the circuit
- B. In parallel with the circuit
- C. In quadrature with the circuit
- D. In phase with the circuit

B   You would usually connect a voltmeter in parallel with the circuit under test.

T8F03 What happens inside a voltmeter when you switch it from a lower to a higher voltage range?

- A. Resistance is added in series with the meter
- B. Resistance is added in parallel with the meter
- C. Resistance is reduced in series with the meter
- D. Resistance is reduced in parallel with the meter

A   The voltmeter is a basic meter movement with a resistor in series. A switch selects an appropriate resistor to set the measurement range of the meter. Higher voltage ranges require higher series resistance values.

T8F04 How is an ammeter usually connected to a circuit under test?

A. In series with the circuit
B. In parallel with the circuit
C. In quadrature with the circuit
D. In phase with the circuit

A   An ammeter is placed in series with the circuit under test. That way, all the current flowing in the circuit must pass through the meter.

T8F05 Which instrument would you use to measure electric current?

A. An ohmmeter
B. A wavemeter
C. A voltmeter
D. An ammeter

D   The unit of electric current is the ampere. You would use an ammeter to measure electric current.

T8F06 What test instrument would be useful to measure DC resistance?

A. An oscilloscope
B. A spectrum analyzer
C. A noise bridge
D. An ohmmeter

D   The unit of resistance is the ohm. You would use an ohmmeter to measure resistance.

T8F07 What might damage a multimeter that uses a moving-needle meter?

A. Measuring a voltage much smaller than the maximum for the chosen scale
B. Leaving the meter in the milliamps position overnight
C. Measuring voltage when using the ohms setting
D. Not allowing it to warm up properly

C   If you have your meter connected in a circuit to make a voltage measurement, with power applied to the circuit, and then switch your meter to a resistance scale, you could burn out the moving-needle meter movement. This would destroy the meter circuitry.

T8F08 For which of the following measurements would you normally use a multimeter?

A. SWR and power
B. Resistance, capacitance and inductance
C. Resistance and reactance
D. Voltage, current and resistance

D   A multimeter is a piece of test equipment that most amateurs should know how to use. The simplest kind of multimeter is the volt-ohm-milliammeter (VOM). As its name implies, a VOM measures voltage, resistance and current. VOMs use one basic meter movement for all functions.

T8F09 What is used to measure relative signal strength in a receiver?

A. An S meter
B. An RST meter
C. A signal deviation meter
D. An SSB meter

A   An S meter monitors a receiver's AGC line to provide an indication of relative signal strength. Your receiver's S meter gives you signal strength comparisons that are far more sensitive and accurate than relying on your ears.

T8F10 With regard to a transmitter and antenna system, what does "forward power" mean?

A. The power traveling from the transmitter to the antenna
B. The power radiated from the top of an antenna system
C. The power produced during the positive half of an RF cycle
D. The power used to drive a linear amplifier

A   The power traveling from the transmitter to the antenna is called forward power. You can measure forward power with a directional wattmeter.

T8F11 With regard to a transmitter and antenna system, what does "reflected power" mean?

A. The power radiated down to the ground from an antenna
B. The power returned towards the source on a transmission line
C. The power produced during the negative half of an RF cycle
D. The power returned to an antenna by buildings and trees

B   When power reaches the antenna in an unmatched system, some of that power is reflected back down the feed line toward the transmitter. (Some of the power is also radiated from the antenna, which is what you want to happen.) The power that returns to the transmitter from the antenna is called reflected power.

T8F12 At what line impedance do most RF wattmeters usually operate?

A. 25 ohms
B. 50 ohms
C. 100 ohms
D. 300 ohms

B   Wattmeters are designed to operate at a certain line impedance, normally 50 ohms. Make sure the feed-line impedance is the same as the design impedance of the wattmeter. If impedances are different, any measurements will be inaccurate.

T8F13 If a directional RF wattmeter reads 90 watts forward power and 10 watts reflected power, what is the actual transmitter output power?

    A. 10 watts
    B. 80 watts
    C. 90 watts
    D. 100 watts

B   To find the true power from your transmitter, subtract the reflected power from the forward reading. (The power reflected from the antenna will again be reflected by the transmitter. This power adds to the forward power reading on the meter.) In this case true power is 90 W (forward) minus 10 W (reflected), which equals 80 W (true power).

T8F14 What is the minimum FCC certification required for an amateur radio operator to build or modify their own transmitting equipment?

    A. A First-Class Radio Repair License
    B. A Technician class license
    C. A General class license
    D. An Amateur Extra class license

B   FCC requires that you have at least a Technician class license if you are going to build or modify your own transmitting equipment.

T8F15 What safety step should you take when soldering?

    A. Always wear safety glasses
    B. Ensure proper ventilation
    C. Make sure no one can touch the soldering iron tip for at least 10 minutes after it is turned off
    D. All of these choices are correct

D   When you are soldering, the soldering iron is hot, so make very sure you don't touch it — let a soldering iron cool down for at least 10 minutes before touching the hot part. You should always wear safety glasses when soldering. Splashes of molten solder can do real damage! Make sure also that the area where you're doing your soldering is well ventilated because fumes from the flux used in soldering can be hazardous.

T8F16 Where would you connect a voltmeter to a 12-volt transceiver if you think the supply voltage may be low when you transmit?

A. At the battery terminals
B. At the fuse block
C. Midway along the 12-volt power supply wire
D. At the 12-volt plug on the chassis of the equipment

D   If you think the supply voltage to your 12-volt transceiver may be low when you transmit, you should check the voltage as it enters the transceiver. The obvious place to do this is at the 12-volt plug on the chassis of the equipment.

T8F17 If your mobile transceiver does not power up, what might you check first?

A. The antenna feedpoint
B. The coaxial cable connector
C. The microphone jack
D. The 12-volt fuses

D   If your mobile transceiver does not power up, you should make sure that it has power. You'll want to check the 12-volt fuses as the first step.

T8F18 What device produces a stable, low-level signal that can be set to a desired frequency?

A. A wavemeter
B. A reflectometer
C. A signal generator
D. An oscilloscope

C   A signal generator is an instrument that can generate RF at various frequencies and at various amplitude levels. This is a very handy instrument for troubleshooting a radio as well as for measuring its performance capabilities — such as the sensitivity of a receiver, the bandwidth of a filter or the ability of a receiver to discriminate against strong adjacent-channel unwanted signals.

T8F19 In Figure T8-9, what circuit quantity would meter B indicate?

    A. The voltage across the resistor
    B. The power consumed by the resistor
    C. The power factor of the resistor
    D. The current flowing through the resistor

D   Meter B is in series with the resistor. That means that all the current flowing through the resistor also flows through the meter. Meter B is an ammeter.

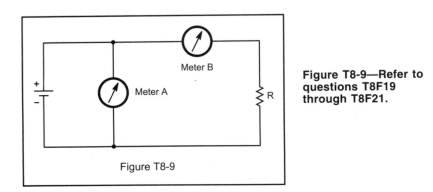

Figure T8-9

Figure T8-9—Refer to questions T8F19 through T8F21.

T8F20 In Figure T8-9, what circuit quantity is meter A reading?

    A. Battery current
    B. Battery voltage
    C. Battery power
    D. Battery current polarity

B   Meter A is connected in parallel with the battery. In that position it will measure the battery voltage. Meter A is a voltmeter.

T8F21 In Figure T8-9, how would the power consumed by the resistor be calculated?

    A. Multiply the value of the resistor times the square of the reading of meter B
    B. Multiply the value of the resistor times the reading of meter B
    C. Multiply the reading of meter A times the value of the resistor
    D. Multiply the value of the resistor times the square root of the reading of meter B

A   Do you remember the power formula? Power equals current squared times resistance. Meter B reads the current, I. Power $= I^2R$.

Special Operations

There will be 2 questions on your exam taken from the Special Operations subelement printed in this chapter. These questions are divided into 2 groups, labeled T9A and T9B.

T9A **How an FM Repeater Works; Repeater operating procedures; Available frequencies; Input/output frequency separation; Repeater ID requirements; Simplex operation; Coordination; Time out; Open/closed repeater; Responsibility for interference.**

T9A01 What is the purpose of repeater operation?
A. To cut your power bill by using someone else's higher power system
B. To help mobile and low-power stations extend their usable range
C. To transmit signals for observing propagation and reception
D. To communicate with stations in services other than amateur

B A repeater receives a signal and retransmits it, usually with higher power, better antennas and from a superior location, to provide an expanded communications range. VHF and UHF repeaters can greatly extend the operating range of amateurs using mobile and hand-held transceivers. If a repeater serves an area, it's not necessary for everyone to live on a hilltop. You only have to be able to hear the repeater's transmitter and reach the repeater's receiver with your transmitted signal. See the drawing on the next page.

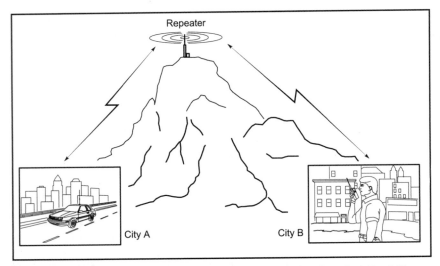

Stations in city A can easily communicate with each other, but the hill blocks their communications with city B. The hilltop repeater enables the groups to communicate with each other.

T9A02 What is a courtesy tone, as used in repeater operations?

- A. A sound used to identify the repeater
- B. A sound used to indicate when a transmission is complete
- C. A sound used to indicate that a message is waiting for someone
- D. A sound used to activate a receiver in case of severe weather

B The courtesy tone used on some repeaters prompts users to leave a space between transmissions. The beeper sounds a second or two after each transmission ends. This permits other stations to transmit their call signs in the intervening time. The conversation continues after the beeper sounds. If a station is too quick and begins transmitting before the beeper sounds, the repeater time-out timer may not reset. In that case the repeater may shut down.

T9A03 During commuting rush hours, which type of repeater operation should be discouraged?

A. Mobile stations
B. Low-power stations
C. Highway traffic information nets
D. Third-party communications nets

D Repeaters were originally intended to enhance mobile communications. During commuter rush hours, mobile stations still have preference over fixed stations on some repeaters. Use good judgment: Rush hours are not the time to test your radio extensively or to join a net that doesn't deal with the weather, highway conditions or other subjects related to commuting. Third-party communications nets also should not be conducted on a repeater during prime commuting hours.

T9A04 Which of the following is a proper way to break into a conversation on a repeater?

A. Wait for the end of a transmission and start calling the desired party
B. Shout, "break, break!" to show that you're eager to join the conversation
C. Turn on an amplifier and override whoever is talking
D. Say your call sign during a break between transmissions

D When you wish to break into a conversation on a repeater, the proper procedure is to give your call sign as soon as one of the stations stops transmitting, while the other station pauses to listen for any breakers. If you are one of the people involved in an ongoing conversation, make sure you briefly pause before you begin each transmission. This allows other stations to break in — there could be an emergency.

T9A05 When using a repeater to communicate, which of the following do you need to know about the repeater?

A. Its input frequency and offset
B. Its call sign
C. Its power level
D. Whether or not it has an autopatch

A To use a repeater, you must have a transceiver that can transmit on the repeater's input frequency and receive on the repeater's output frequency. The input and output frequencies are separated by a predetermined amount called the offset (or split). To operate through a certain repeater, you'll need to know the repeater's input (or output) frequency and the offset.

T9A06 Why should you pause briefly between transmissions when using a repeater?

    A. To check the SWR of the repeater
    B. To reach for pencil and paper for third-party communications
    C. To listen for anyone wanting to break in
    D. To dial up the repeater's autopatch

C   If you are involved in an ongoing conversation, make sure you briefly pause before you begin each transmission. This allows other stations to break in — there could be an emergency. Don't key your microphone as soon as someone else releases theirs. If your exchanges are too quick, you can prevent other stations from getting in.

T9A07 Why should you keep transmissions short when using a repeater?

    A. A long transmission may prevent someone with an emergency from using the repeater
    B. To see if the receiving station operator is still awake
    C. To give any listening non-hams a chance to respond
    D. To keep long-distance charges down

A   Keep transmissions as short as possible, so more people can use the repeater. Even more important, long transmissions could prevent someone with an emergency from getting the chance to call for help through the repeater. All repeaters encourage short transmissions by "timing out" (shutting down for a few minutes) when someone gets longwinded.

T9A08 How could you determine if a repeater is already being used by other stations?

    A. Ask if the frequency is in use, then give your call sign
    B. If you don't hear anyone, assume that the frequency is clear to use
    C. Check for the presence of the CTCSS tone
    D. If the repeater identifies when you key your transmitter, it probably was already in use

A   One way you could determine if other stations are already using a repeater is to ask if the frequency is in use, then give your call sign. Don't forget to give your call sign. Proper identification is required by the Rules.

T9A09 What is the usual input/output frequency separation for repeaters in the 2-meter band?

A. 600 kHz
B. 1.0 MHz
C. 1.6 MHz
D. 5.0 MHz

A On the 2-meter (144 to 148 MHz) band, most repeaters use an input/output frequency separation of 600 kHz. **Table 9-1** summarizes the standard frequency offset between the repeater input and output frequencies on various bands.

Table 9-1

Repeater Input/Output Offsets

Band	Offset
6 meters	1 MHz
2 meters	600 kHz
1.25 meters	1.6 MHz
70 cm	5 MHz
33 cm	12 MHz
23 cm	20 MHz

T9A10 What is the usual input/output frequency separation for repeaters in the 70-centimeter band?

A. 600 kHz
B. 1.0 MHz
C. 1.6 MHz
D. 5.0 MHz

D On the 70-cm (420 to 450 MHz) band the standard offset is 5.0 MHz. You can see in **Table 9-1** that where there is more space available on the band, a wider offset is chosen. By providing more space between the input and output frequencies there is less chance for interference or interaction between the two.

T9A11 What does it mean to say that a repeater has an input and an output frequency?

A. The repeater receives on one frequency and transmits on another
B. The repeater offers a choice of operating frequency, in case one is busy
C. One frequency is used to control the repeater and another is used to retransmit received signals
D. The repeater must receive an access code on one frequency before retransmitting received signals

A   A repeater receives a signal on one frequency and simultaneously retransmits (repeats) it on another frequency. The frequency it receives on is called the input frequency, and the frequency it transmits on is called the output frequency.

T9A12 What is the most likely reason you might hear Morse code tones on a repeater frequency?

A. Intermodulation
B. An emergency request for help
C. The repeater's identification
D. A courtesy tone

C   Many repeaters identify themselves periodically using Morse code. That is the most likely reason that you might hear Morse code being sent on a repeater.

T9A13 What is the common amateur meaning of the term "simplex operation"?

A. Transmitting and receiving on the same frequency
B. Transmitting and receiving over a wide area
C. Transmitting on one frequency and receiving on another
D. Transmitting one-way communications

A   Simplex operation means the stations are talking to each other by transmitting and receiving on the same frequency. This is similar to making a contact on the HF bands.

T9A14 When should you use simplex operation instead of a repeater?

    A. When the most reliable communications are needed
    B. When a contact is possible without using a repeater
    C. When an emergency telephone call is needed
    D. When you are traveling and need some local information

B   Use simplex whenever you can make the contact without using a repeater. The repeater is not a soapbox. You may like to listen to yourself, but others who may need to use the repeater will not appreciate your tying up the repeater unnecessarily.

T9A15 If you are talking to a station using a repeater, how would you find out if you could communicate using simplex instead?

    A. See if you can clearly receive the station on the repeater's input frequency
    B. See if you can clearly receive the station on a lower frequency band
    C. See if you can clearly receive a more distant repeater
    D. See if a third station can clearly receive both of you

A   The easiest way to determine if you are able to communicate with the other station on simplex is to listen to the repeater input frequency. Since this is the frequency the other station uses to transmit to the repeater, if you can hear his signals there, you should be able to use simplex.

T9A16 What is it called if the frequency coordinator recommends that you operate on a specific repeater frequency pair?

    A. FCC type acceptance
    B. FCC type approval
    C. Frequency division multiplexing
    D. Repeater frequency coordination

D   Volunteer frequency coordinators have been appointed to ensure that new repeaters use frequencies that will tend not to interfere with existing repeaters in the same area. The process is called repeater frequency coordination. The FCC encourages frequency coordination, but the process is organized and run by hams and groups of hams who use repeaters.

T9A17 What is the purpose of a repeater time-out timer?

    A. It lets a repeater have a rest period after heavy use

    B. It logs repeater transmit time to predict when a repeater will fail

    C. It tells how long someone has been using a repeater

    D. It limits the amount of time a repeater can transmit continuously

D   A time-out timer prevents a repeater from transmitting continuously, due to distant signals or interference. Because it has such a wide coverage area, a continuously transmitting repeater could cause unnecessary interference. Continuous operation can also damage the repeater.

T9A18 What should you do if you hear a closed repeater system that you would like to be able to use?

    A. Contact the control operator and ask to join

    B. Use the repeater until told not to

    C. Use simplex on the repeater input until told not to

    D. Write the FCC and report the closed condition

A   Some repeaters have limited access. The owner or owners may decide to restrict access to a small group, perhaps just the members of a club. If you wish to join a group that sponsors such a closed repeater, contact the repeater's control operator or sponsoring club.

T9A19 Who pays for the site rental and upkeep of most repeaters?

    A. All amateurs, because part of the amateur license examination fee is used

    B. The repeater owner and donations from its users

    C. The Federal Communications Commission

    D. The federal government, using money granted by Congress

B   A repeater site may well be rented, and the repeater hardware and antenna are usually funded and maintained by either an individual or a club on a voluntary basis. Repeater owners usually accept and frequently encourage donations from users to pay operational expenses.

T9A20 If a repeater is causing harmful interference to another amateur repeater and a frequency coordinator has recommended the operation of both repeaters, who is responsible for resolving the interference?

A. The licensee of the repeater that has been recommended for the longest period of time

B. The licensee of the repeater that has been recommended the most recently

C. The frequency coordinator

D. Both repeater licensees

D   The FCC has ruled in favor of coordinated repeaters if there is harmful interference between two repeaters. In such a case, if a frequency coordinator has coordinated one but not the other, the licensee of the uncoordinated repeater is responsible for solving the interference problem. If both repeaters are coordinated, then both licensees are equally responsible for resolving the interference. [97.205 (c)]

T9B   Beacon, satellite, space, EME communications; Radio control of models; Autopatch; Slow scan television; Telecommand; CTCSS tone access; Duplex/crossband operation.

T9B01 What is an amateur station called that transmits communications for the purpose of observation of propagation and reception?

A. A beacon

B. A repeater

C. An auxiliary station

D. A radio control station

A   A beacon station is simply a transmitter that alerts listeners to its presence. In the amateur service, beacons are used primarily for the study of radiowave propagation — to allow amateurs to tell when a band is open to different parts of the country or world. The FCC defines a beacon station as an amateur station transmitting communications for the purposes of observation of propagation and reception or other related experimental activities. [97.3 (a) (9)]

T9B02 Which of the following is true of amateur radio beacon stations?

A. Automatic control is allowed in certain band segments
B. One-way transmissions are permitted
C. Maximum output power is 100 watts
D. All of these choices are correct

D   The FCC Rules allow one-way transmissions by beacon stations. Further, automatically controlled beacon stations are allowed in certain parts of the 28, 50, 144, 222 and 432-MHz amateur bands, and all amateur bands above 450 MHz. However, the transmitter power of a beacon must not exceed 100 W. [97.203 (c), (d), (g)]

T9B03 The control operator of a station communicating through an amateur satellite must hold what class of license?

A. Amateur Extra or Advanced
B. Any class except Novice
C. Any class
D. Technician with satellite endorsement

C   The control operator of a station communicating through an amateur satellite may hold any class of amateur license. If you have transmitting privileges on the satellite input (receive) frequency range, you may communicate through that satellite. The FCC does limit the use of satellites based on license class. [97.209 (a)]

T9B04 How does the Doppler effect change an amateur satellite's signal as the satellite passes overhead?

A. The signal's amplitude increases or decreases
B. The signal's frequency increases or decreases
C. The signal's polarization changes from horizontal to vertical
D. The signal's circular polarization rotates

B   Doppler shift or Doppler effect is the name given to describe the way the downlink frequency of a satellite varies by several kilohertz during a low-earth orbit. Doppler shift is caused by the relative motion between you and the satellite. In operation, as the satellite is moving toward you, the frequency of a downlink signal appears to be increased by a small amount. When the satellite passes overhead and begins to move away from you, there will be a sudden frequency drop of a few kilohertz, in much the same was as the tone of a car horn or train whistle drops as the vehicle moves past you.

T9B05 Why do many amateur satellites operate on the VHF/UHF bands?

A. To take advantage of the skip zone
B. Because VHF/UHF equipment costs less than HF equipment
C. To give Technician class operators greater access to modern communications technology
D. Because VHF and UHF signals easily pass through the ionosphere

D   Amateur Radio operators have built many satellites since the first one was launched in 1961. The satellites often use the VHF and UHF bands because radio signals on those bands normally go right through the ionosphere. The satellites retransmit signals to provide greater communications range than would normally be possible on those bands. While it is possible to use HF signals for satellite operation (and some satellites do), more of the HF signal energy may be bent back to the Earth rather than going through to the satellite.

T9B06 Which antenna system would NOT be a good choice for an EME (moonbounce) station?

A. A parabolic-dish antenna
B. A multi-element array of collinear antennas
C. A ground-plane antenna
D. A high-gain array of Yagi antennas

C   The Moon's average distance from the Earth is 239,000 miles, and EME signals must travel twice the distance to the Moon. The total signal loss between the transmitting and receiving stations are huge when compared to local VHF paths. A typical EME station uses high-gain antennas and a high-power amplifier. For example, a high-gain array of Yagi antennas would be an appropriate choice for a moonbounce station. You would not even want to try moonbounce with a simple ground-plane antenna, no matter how much transmitter power you had!

T9B07 What does the term "apogee" refer to when applied to an Earth satellite?

    A. The closest point to the Earth in the satellite's orbit
    B. The most distant point from the Earth in the satellite's orbit
    C. The point where the satellite appears to cross the equator
    D. The point when the Earth eclipses the satellite from the sun

B   The farthest distance from the center of the Earth in a satellite's orbit is called the apogee. The first part of apogee comes from the same word as apex, which means the high point. The second part derives from "geo," which relates to the Earth.

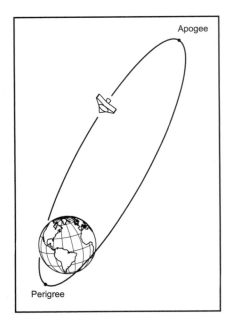

Apogee

Perigree

Apogee is that point in a satellite's orbit when it is farthest from the Earth. Perigee is the point in the orbit when it is closest to the Earth.

T9B08 What does the term "perigee" refer to when applied to an Earth satellite?

    A. The closest point to the Earth in the satellite's orbit
    B. The most distant point from the Earth in the satellite's orbit
    C. The time when the satellite will be on the opposite side of the Earth
    D. The effect that causes the satellite's signal frequency to change

A   A satellite's orbit around the Earth is usually shaped like an ellipse. In fact, a perfectly circular orbit is one kind of ellipse. Most orbits, however, are not perfectly circular. The distance from the center of the Earth to a satellite's closest approach to the Earth is called the perigee.

T9B09 What mathematical parameters describe a satellite's orbit?

A. Its telemetry data
B. Its Doppler shift characteristics
C. Its mean motion
D. Its Keplerian elements

D   Johannes Kepler described the planetary orbits of our solar system. The laws and mathematical formulas that he developed may be used to calculate the location of a satellite at any given time. If you know the values of a set of measurements about the satellite orbit, called "Keplerian elements," you can do the calculations. These days, most folks use computer software to do the math.

T9B10 What is the typical amount of time an amateur has to communicate with the International Space Station?

A. 4 to 6 minutes per pass
B. An hour or two per pass
C. About 20 minutes per pass
D. All day

A   Contacts from a well-equipped ground station with the International Space Station (ISS) — which is really just a huge satellite — typically last from 4 to 6 minutes for each pass, since it has a fairly low, nearly circular orbit with a period of about 90 minutes.

T9B11 Which of the following would be the best emission mode for two-way EME contacts?

A. CW
B. AM
C. FM
D. Spread spectrum

A   The Moon's average distance from the Earth is 239,000 miles, and EME signals must travel twice the distance to the Moon. Path losses are huge when compared to "local" VHF paths. In addition to the long distance to the Moon and back, the Moon's surface is irregular and not a particularly efficient reflector of radio waves. Because signals received from the Moon are very weak, most operators use CW, with narrow receiving filters.

T9B12 What minimum information must be on a label affixed to a transmitter used for telecommand (control) of model craft?

A. Station call sign
B. Station call sign and the station licensee's name
C. Station call sign and the station licensee's name and address
D. Station call sign and the station licensee's class of license

C   The FCC Rules for telecommand operation of a model craft don't require that identification be used, since identification is impractical in that application. However, there must be an identification label with the station call sign and the licensee's name and address affixed to the transmitter. [97.215 (a)]

T9B13 What is an autopatch?

A. An automatic digital connection between a US and a foreign amateur
B. A digital connection used to transfer data between a hand-held radio and a computer
C. A device that allows radio users to access the public telephone system
D. A video interface allowing images to be patched into a digital data stream

C   An autopatch allows repeater users to make telephone calls through the repeater. To use most repeater autopatches, you generate the standard telephone company tones to access and dial through the system. The tones are usually generated with a telephone-type tone pad connected to the transceiver.

T9B14 Which of the following statements about Amateur Radio autopatch usage is true?

A. The person called using the autopatch must be a licensed radio amateur
B. The autopatch will allow only local calls to police, fire and ambulance services
C. Communication through the autopatch is not private
D. The autopatch should not be used for reporting emergencies

C   Any telephone conversation that you have using the autopatch feature of a repeater will be transmitted through the repeater. That means that anyone listening on the repeater transmit frequency will hear your conversation. In other words, your communications through the autopatch are not private.

T9B15 Which of the following will allow you to monitor Amateur Television (ATV) on the 70-cm band?

A. A portable video camera
B. A cable ready TV receiver
C. An SSTV converter
D. A TV flyback transformer

B   Several cable-TV channels operate at frequencies in the Amateur 420 to 450 MHz band. Cable channels 57 through 61 are in that band. (Note: These are cable channels, not broadcast channels.) So a cable-ready TV set to receive one of those channels should be able to pick up any local ATV activity when used with a good outdoor antenna — not your Cable TV feed line!

T9B16 When may slow-scan television be transmitted through a 2-meter repeater?

A. At any time, providing the repeater control operator authorizes this unique transmission
B. Never; slow-scan television is not allowed on 2 meters
C. Only after 5:00 PM local time
D. Never; slow-scan television is not allowed on repeaters

A   Slow-Scan Television (SSTV) is popular on 20-meter HF using SSB, and it is also used on VHF/UHF with FM radios. Some 2-meter voice repeaters also allow SSTV operation on their machines. It can take almost two minutes to transmit a color SSTV picture. Also, listeners will hear tones—not voices during the picture transmission. For those reasons you should have authorization from the repeater control operator before you transmit SSTV through a 2-meter repeater.

T9B17 What is the definition of telecommand?

    A. All communications using the telephone or telegraphy with space stations

    B. A one way transmission to initiate conversation with astronauts aboard a satellite or space station

    C. A one way transmission to initiate, modify or terminate functions of a device at a distance

    D. Two way transmissions to initiate, modify or terminate functions of a device at a distance

C   Telecommand means literally "command from a distance." Telecommand operation is a one-way transmission to initiate, modify or terminate functions of a device at a distance. A familiar example is remote control of a model aircraft or boat, but telecommand operation can also include control of an amateur satellite or a remotely located amateur station. [97.3 (a) (43)]

T9B18 What provisions must be in place for the legal operation of a telecommand station?

    A. The station must have a wire line or radio control link

    B. A photocopy of the station license must be posted in a conspicuous location

    C. The station must be protected so that no unauthorized transmission can be made

    D. All of these choices are correct

D   Each one of the choices is a requirement under FCC Rules. In other words, all of these choices are correct. [97.213 (a), (b), (c)]

T9B19 What is a continuous tone-coded squelch system (CTCSS) tone (sometimes called PL — a Motorola trademark)?

A. A special signal used for telecommand control of model craft
B. A sub-audible tone, added to a carrier, which may cause a receiver to accept the signal
C. A tone used by repeaters to mark the end of a transmission
D. A special signal used for telemetry between amateur space stations and Earth stations

B   CTCSS adds a sub-audible tone to a transmitted carrier. A CTCSS receiver requires that the proper tone be present for the squelch to open and the receiver to accept the signal. CTCSS is used to reduce co-channel interference during band openings. Without it repeaters can be activated by distant signals on their input frequencies during band openings. That's a situation that only adds to congestion and frustration.

T9B20 What does it mean if you are told that a tone is required to access a repeater?

A. You must use keypad tones like your phone system to operate it
B. You must wait to hear a warbling two-tone signal to operate it
C. You must wait to hear a courtesy beep tone at the end of another's transmission before you can operate it
D. You must use a subaudible tone-coded squelch with your signal to operate it

D   If a repeater is equipped with CTCSS, you'll need to transmit the correct tone along with your voice signal. The tone is required to open the receiver squelch and thus activate (operate) that repeater.

T9B21 What is the term that describes a repeater that receives signals on one band and retransmits them on another band?

A. A special coordinated repeater
B. An illegally operating repeater
C. An auxiliary station
D. A crossband repeater

D   Some specialized repeaters receive on one amateur band and retransmit on another amateur band. These are called crossband repeaters. Because they use widely separated frequencies, crossband repeaters can even be used for full-duplex operation, much like a regular telephone allows the user to talk and listen simultaneously. A crossband repeater might receive on the 70-cm band and retransmit on the 2-meter band, for example.

Electrical, Antenna Structure and RF Safety Practices

There will be 6 questions on your exam taken from the Electrical, Antenna Structure and RF Safety Practices subelement printed in this chapter. These questions are divided into 6 groups, labeled T0A through T0F.

T0A **Sources of electrical danger in amateur stations: lethal voltages, high current sources, fire; avoiding electrical shock; Station wiring; Wiring a three wire electrical plug; Need for main power switch; Safety interlock switch; Open/short circuit; Fuses; Station grounding.**

T0A01 What is the minimum voltage that is usually dangerous to humans?

- A. 30 volts
- B. 100 volts
- C. 1000 volts
- D. 2000 volts

A Low-voltage power supplies may seem safe, but even battery-powered equipment should be treated with care. The minimum voltage considered dangerous to humans is 30 volts. You should respect even low voltages, taking appropriate steps to avoid contact.

TOA02 Which electrical circuit draws high current?

    A. An open circuit
    B. A dead circuit
    C. A closed circuit
    D. A short circuit

D   You've probably heard the term short circuit before. A short circuit happens when the current flowing through the components doesn't follow the path one would expect it to. Instead, the current finds another path — a shorter one — between the terminals of the power source. This is why they call this path a short circuit. There is less opposition to the flow of electrons, so there is a larger current. Often the current through the new (short) path is so large that the wires or components can't handle it. When this happens, the wires and components can be damaged.

TOA03 What could happen to your transceiver if you replace its blown 5 amp AC line fuse with a 30 amp fuse?

    A. The 30-amp fuse would better protect your transceiver from using too much current
    B. The transceiver would run cooler
    C. The transceiver could use more current than 5 amps and a fire could occur
    D. The transceiver would not be able to produce as much RF output

C   Do not put a larger fuse in an existing circuit — too much current could be drawn. The wires would become hot and a fire could result. If your transceiver blows a fuse in the main ac power line, you should find out what caused the fuse to blow and repair the problem. Replace the fuse only with another one of the same rating. For example if you replaced a 5-amp fuse with one rated for 30 amps the transceiver could draw too much current, causing the wires to overheat and even starting a fire.

TOA04 How much electrical current flowing through the human body will probably be fatal?

    A. As little as 1/10 of an ampere
    B. Approximately 10 amperes
    C. More than 20 amperes
    D. Current through the human body is never fatal

A   You should never underestimate the potential hazard when working with electricity. As little as 100 milliamps (mA), or 1/10 amp (A), can be fatal! As the saying goes, "It's volts that jolts, but it's mills that kills." Even battery-powered equipment should be treated with respect. Automobile batteries are designed to provide very high current (as much as 200 A) for short periods when starting a car. This much current can kill you, even at 12 volts. See **Table 10-1**.

Table 10-1

Effects of Electric Current Through the Body of an Average Person

Current (1 Second Contact)	Effect
1 mA	Just Perceptible.
5 mA	Maximum harmless current.
10 - 20 mA	Lower limit for sustained muscular contractions.
30 - 50 mA	Pain
50 mA	Pain, possible fainting. "Can't let go" current.
100 - 300 mA	Normal heart rhythm disrupted. Electrocution if sustained current.
6 A	Sustained heart contractions. Burns if current density is high.

TOA05 Which body organ can be fatally affected by a very small amount of electrical current?

    A. The heart
    B. The brain
    C. The liver
    D. The lungs

A   A small electrical current can upset the operation of your heart. Even a very low current can cause heart failure and death.

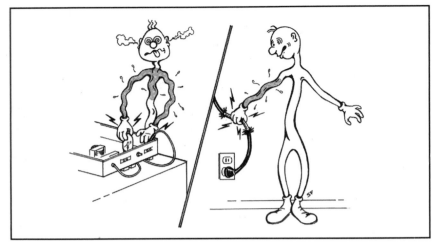

The path from the electrical source to ground affects how severe an electrical shock will be. The most dangerous path (from hand to hand directly through the heart) is shown at the left. The path from one finger to the other shown at the right is not quite so dangerous.

T0A06 For best protection from electrical shock, what should be grounded in an amateur station?

A. The power supply primary
B. All station equipment connected to a common ground
C. The antenna feed line
D. The AC power mains

B   The best way to protect against electrical shock is to make sure that each part of your station is at the same potential. That potential should, of course, be that of ground. You'll want to make sure that you have a common ground. That means that all the grounds are tied together.

T0A07 Which potential does the green wire in a three-wire electrical plug represent?

A. Neutral
B. Hot
C. Hot and neutral
D. Ground

D   State and national electrical-safety codes require three-wire power cords on many 120-V tools and appliances. Power supplies and station equipment use similar connections. Two of the conductors (the "hot" and "neutral" wires) power the device. The third conductor (the safety ground wire) connects to the metal frame of the device. The "hot" wire is usually black or red. The "neutral" wire is white. The frame/ground wire is green or sometimes bare.

T0A08 What is an important consideration for the location of the main power switch?

A. It must always be near the operator
B. It must always be as far away from the operator as possible
C. Everyone should know where it is located in case of an emergency
D. It should be located in a locked metal box so no one can accidentally turn it off

C   If you should discover someone who is being burned by high voltage, immediately turn off the power, call for help and give cardiopulmonary resuscitation (CPR). Be sure all family members know how to turn off power at the main power switch. These measures could save your life or that of a friend or family member.

T0A09 What circuit should be controlled by a safety interlock switch in an amateur transceiver or power amplifier?

A. The power supply
B. The IF amplifier
C. The audio amplifier
D. The cathode bypass circuit

A   There should be a device that turns power off automatically if you remove covers or shielding. Such a safety interlock reduces the danger of electrical shock from high voltages when you open the cabinet. The safety interlock switch should control the voltage applied to the power supply circuit. Any equipment that connects to the 120-V or 240-V ac supply should include such a safety switch.

T0A10 What type of electrical circuit is created when a fuse blows?

    A. A closed circuit
    B. A bypass circuit
    C. An open circuit
    D. A short circuit

    C   When a fuse blows or a circuit breaker trips it creates an open circuit. We use fuses or circuit breakers in our house wiring and in electrical equipment to protect against the large current drawn by a short circuit or an overloaded circuit.

T0A11 Why would it be unwise to touch an ungrounded terminal of a high voltage capacitor even if it's not in an energized circuit?

    A. You could damage the capacitor's dielectric material
    B. A residual charge on the capacitor could cause interference to others
    C. You could damage the capacitor by causing an electrostatic discharge
    D. You could receive a shock from a residual stored charge

    D   A capacitor stores electrical energy. Even after you shut off a circuit, a high voltage capacitor may retain a potentially harmful charge.

T0A12 What safety equipment item should you always add to home built equipment that is powered from 110 volt AC lines?

    A. A fuse or circuit breaker in series with the equipment
    B. A fuse or circuit breaker in parallel with the equipment
    C. Install Zener diodes across AC inputs
    D. House the equipment in a plastic or other non-conductive enclosure

    A   To protect against unexpected short circuits and other problems, most electronic equipment includes one or more fuses. When the fuse (usually placed in the main power line to the equipment) blows, it creates an open circuit, stopping the current. You should include a fuse or circuit breaker in series with the 110 volt ac lines to any home-built equipment that you might have.

TOA13 When fuses are installed in 12-volt DC wiring, where should they be placed?

    A. At the radio
    B. Midway between voltage source and radio
    C. Fuses aren't required for 12-volt DC equipment
    D. At the voltage source

    D   Fuses must be placed between the voltage source and the circuit that you want to protect. If you place the fuses in a 12-volt dc wiring setup at the voltage source, then the entire circuit beyond the fuses will be protected. That means not just your transceiver is protected from shorts and overloads, but so also are the wires that conduct the 12 volts from the fuses to the transceiver.

T0B   **Lightning protection; Antenna structure installation safety; Tower climbing Safety; Safety belt/hard hat/ safety glasses; Antenna structure limitations.**

T0B01 How can an antenna system best be protected from lightning damage?

    A. Install a balun at the antenna feed point
    B. Install an RF choke in the antenna feed line
    C. Ground all antennas when they are not in use
    D. Install a fuse in the antenna feed line

    C   The best protection against lightning is to disconnect all antennas and rotator control cables from your equipment and connect them to ground. You should also form the habit of unplugging all power cords when you aren't on the air. It takes time to hook up everything when you want to operate again, but you will protect your station and your home if you follow this simple precaution.

T0B02 How can amateur station equipment best be protected from lightning damage?

    A. Use heavy insulation on the wiring
    B. Never turn off the equipment
    C. Disconnect the ground system from all radios
    D. Disconnect all equipment from the power lines and antenna cables

D   When your station is not being used, you should disconnect and ground all antennas, feed lines and rotator cables for effective lightning protection. An ungrounded antenna can pick up large electrical charges from storms in the area. These charges can damage your equipment (particularly receivers) if you don't take precautions.

You should also unplug your equipment. Why unplug your equipment if the antennas are disconnected? Lightning can still find its way into your equipment through the power cord. Power lines can act as long antennas, picking up sizable charges during a storm. Simply turning off the main circuit breaker is not enough — lightning can easily jump over the circuit breaker contacts and find its way into your equipment.

T0B03 Why should you wear a hard hat and safety glasses if you are on the ground helping someone work on an antenna tower?

    A. So you won't be hurt if the tower should accidentally fall
    B. To keep RF energy away from your head during antenna testing
    C. To protect your head from something dropped from the tower
    D. So someone passing by will know that work is being done on the tower and will stay away

C   Helpers on the ground should never stand directly under the work being done. All ground helpers should wear hard hats and safety glasses for protection. Even a small tool can make quite a dent if it falls from 50 or 60 feet.

TOB04 What safety factors must you consider when using a bow and arrow or slingshot and weight to shoot an antenna-support line over a tree?

A. You must ensure that the line is strong enough to withstand the shock of shooting the weight

B. You must ensure that the arrow or weight has a safe flight path if the line breaks

C. You must ensure that the bow and arrow or slingshot is in good working condition

D. All of these choices are correct

D To prevent personal injury and avoid property damage, you should observe all of the given factors. Safety is of utmost importance. Never take chances or make compromises where safety is an issue.

TOB05 Which of the following is the best way to install your antenna in relation to overhead electric power lines?

A. Always be sure your antenna wire is higher than the power line, and crosses it at a 90-degree angle

B. Always be sure your antenna and feed line are well clear of any power lines

C. Always be sure your antenna is lower than the power line, and crosses it at a small angle

D. Only use vertical antennas within 100 feet of a power line

B Never put your antenna or feed line under, or over the top of electrical power lines. Never place a vertical antenna where it could fall against the electrical power lines. If power lines ever come into contact with your antenna, you could be electrocuted. The only safe place to install an antenna is in a location that is well clear of any power lines.

T0B06 What should you always do before attempting to climb an antenna tower?

    A. Turn on all radio transmitters that use the tower's antennas
    B. Remove all tower grounding to guard against static electric shock
    C. Put on your safety belt and safety glasses
    D. Inform the FAA and the FCC that you are starting work on a tower

C   Antenna work sometimes requires that someone climb up on a tower. Never work alone! Work slowly, thinking out each move before you make it. The person on the tower should always wear a safety belt, and keep it securely anchored. Before each use, inspect the belt carefully for damage such as cuts or worn areas. The belt will make it much easier to work on the antenna and will also prevent an accidental fall. A hard hat and safety glasses are also important safety equipment.

T0B07 What is the most important safety precaution to take when putting up an antenna tower?

    A. Install steps on your tower for safe climbing
    B. Insulate the base of the tower to avoid lightning strikes
    C. Ground the base of the tower to avoid lightning strikes
    D. Look for and stay clear of any overhead electrical wires

D   If power lines ever come into contact with your antenna, you could be electrocuted. The only safe place to install an antenna tower is in a location that is well clear of any power lines. Before you put up a tower, look for any overhead electrical wires. Make sure that the tower is installed where there is no possibility of contact between the lines and the tower if the lines should ever break or the tower should ever fall.

TOB08 What should you consider before you climb a tower with a leather climbing belt?

    A. If the leather is old, it is probably brittle and could break unexpectedly

    B. If the leather is old, it is very tough and is not likely to break easily

    C. If the leather is old, it is flexible and will hold you more comfortably

    D. An unbroken old leather belt has proven its holding strength over the years

A   Before each use, inspect your safety belt carefully for damage such as cuts or worn areas. If you have and old leather belt, it is safer to replace it with a newer fabric model. Old leather becomes brittle and can break unexpectedly. The results can be disastrous. Never take a chance on an old leather climbing belt!

TOB09 What should you do before you climb a guyed tower?

    A. Tell someone that you will be up on the tower

    B. Inspect the tower for cracks or loose bolts

    C. Inspect the guy wires for frayed cable, loose cable clamps, loose turnbuckles or loose guy anchors

    D. All of these choices are correct

D   Before you climb a tower you should tell someone that you will be up on the tower. Make sure that you can get help quickly if it is needed. Before you take the first step up a tower, inspect the tower for cracks and loose bolts. Examine the tower from two or three different angles and use binoculars if you have them available. You'll also need to inspect the guy wires for frayed cable, loose cable clamps, loose turnbuckles or loose guy anchors. Each of these items represents critical safety issues. Make every item a part of your regular routine before each climb.

TOB10 What should you do before you do any work on top of your tower?

A. Tell someone that you will be up on the tower
B. Bring a variety of tools with you to minimize your trips up and down the tower
C. Inspect the tower before climbing to become aware of any antennas or other obstacles that you may need to step around
D. All of these choices are correct

D Before you climb a tower you should always tell someone that you will be up on the tower. Ideally, that person would keep you in sight at all times. Before you begin your climb, inspect the tower not only for safety issues but so that you are aware of any antennas or other obstacles that you may need to step around. Take the tools you'll need with you when you climb, or pull them up on a rope once you've reached the top. A canvas bucket or fabric bag that can be fastened to the tower will hold the tools without burdening you as you work. It doesn't hurt to have an extra adjustable wrench and a pair of locking pliers. Short lengths of rope can be used to secure items while you are working.

TOC   Definition of RF radiation; Procedures for RF environmental safety; Definitions and guidelines.

TOC01 What is radio frequency radiation?

A. Waves of electric and magnetic energy between 3 kHz and 300 GHz
B. Ultra-violet rays emitted by the sun between 20 Hz and 300 GHz
C. Sound energy given off by a radio receiver
D. Beams of X-Rays and Gamma rays emitted by a radio transmitter

A The combination of electric and magnetic waves of energy that are produced by your transmitter are called electromagnetic radiation. Electromagnetic radiation with frequencies between about 3 kHz and 300 GHz is called radio frequency (RF) radiation. (Higher-frequency electromagnetic waves, such as light, ultraviolet radiation and even gamma rays and X-rays are all above the RF range.)

TOC02 Why is it a good idea to adhere to the FCC's Rules for using the minimum power needed when you are transmitting with your hand-held radio?

A. Large fines are always imposed on operators violating this rule
B. To reduce the level of RF radiation exposure to the operator's head
C. To reduce calcification of the NiCd battery pack
D. To eliminate self-oscillation in the receiver RF amplifier

B Higher transmitter power will produce stronger radiated RF fields. So using the minimum power necessary to carry out your communications will minimize the exposure of anyone near your station. When you are using your hand-held radio, using lower power will reduce the level of RF radiation exposure to your head.

TOC03 Which of the following units of measurement are used to specify the power density of a radiated RF signal?

A. Milliwatts per square centimeter
B. Volts per meter
C. Amperes per meter
D. All of these choices are correct

A The basic unit of power is the watt. Density is a measure of how many watts you'll find in a given area. Based on that, you might expect that power density is measured in watts per square meter. If you did, you'd be close but not entirely correct. Power density is usually measured in milliwatts per square centimeter.

TOC04 Over what frequency range are the FCC Regulations most stringent for RF radiation exposure?

A. Frequencies below 300 kHz
B. Frequencies between 300 kHz and 3 MHz
C. Frequencies between 3 MHz and 30 MHz
D. Frequencies between 30 MHz and 300 MHz

D FCC Regulations are most stringent for RF radiation exposure at VHF. That is the frequency range between 30 MHz and 300 MHz. Frequencies in the VHF range include natural resonant frequencies for humans. Body size helps determine the frequency at which RF energy is most easily absorbed.

TOC05 Which of the following categories describes most common amateur use of a hand-held transceiver?

   A. Mobile devices
   B. Portable devices
   C. Fixed devices
   D. None of these choices is correct

  B   Hand-held radios are very popular for VHF and UHF operation, especially with FM repeaters. They transmit with less than 7 watts of power, which is generally considered safe. Because the radios are designed to be operated with an antenna that is within 20 centimeters of your body, they are classified as portable devices by the FCC.

TOC06 From an RF safety standpoint, what impact does the duty cycle have on the minimum safe distance separating an antenna and people in the neighboring environment?

   A. The lower the duty cycle, the shorter the compliance distance
   B. The compliance distance is increased with an increase in the duty cycle
   C. Lower duty cycles subject people in the environment to lower radio-frequency radiation
   D. All of these answers are correct

  D   An emission with a lower duty cycle produces less RF radiation exposure for the same PEP output. Lower duty cycles, then, result in lower RF radiation exposures. That also means the antenna can be closer to people without exceeding their MPE (maximum permissible exposure) limit. Higher duty cycles require a greater distance between your antenna and people to achieve compliance. To sum it up, all of these answers are correct.

TOC07 Why is the concept of "duty cycle" one factor used to determine safe RF radiation exposure levels?

    A. It takes into account the amount of time the transmitter is operating at full power during a single transmission

    B. It takes into account the transmitter power supply rating

    C. It takes into account the antenna feed line loss

    D. It takes into account the thermal effects of the final amplifier

A   The duty cycle of an emission takes into account the amount of time a transmitter is operating at full power during a single transmission. An emission with a lower duty cycle produces less RF radiation exposure for the same PEP output.

TOC08 What factors affect the resulting RF fields emitted by an amateur transceiver that expose people in the environment?

    A. Frequency and power level of the RF field

    B. Antenna height and distance from the antenna to a person

    C. Radiation pattern of the antenna

    D. All of these answers are correct

D   All of the choices are correct. The human body absorbs less RF energy at some frequencies and more at others. If you decrease your transmitter output power, you decrease the RF field radiated from your antenna. You can place your antenna farther from people to lessen their exposure to RF fields. Finally, the radiation pattern of your antenna affects the resulting RF fields radiated.

TOC09 What unit of measurement specifies RF electric field strength?

    A. Coulombs (C) at one wavelength from the antenna

    B. Volts per meter (V/m)

    C. Microfarads (uF) at the transmitter output

    D. Microhenrys (uH) per square centimeter

B   The FCC maximum permissible exposure limits are given in terms of electric and magnetic field strengths. RF electric field strength is measured in volts per meter (V/m)

TOC10 Which of the following is considered to be non-Ionizing radiation?

- A. X-radiation
- B. Gamma radiation
- C. Ultra violet radiation
- D. Radio frequency radiation

D   Both RF and 60-Hz fields are classified as non-ionizing radiation because the frequency is too low for there to be enough photon energy to ionize atoms. Ionizing radiation, such as X-rays, gamma rays and even some ultraviolet radiation has enough energy to knock electrons loose from their atoms. When this happens, positive and negative ions are formed.

TOC11 What do the FCC RF radiation exposure regulations establish?

- A. Maximum radiated field strength
- B. Minimum permissible HF antenna height
- C. Maximum permissible exposure limits
- D. All of these choices are correct

C   Amateur Radio is basically a safe activity. In recent years, however, there has been considerable discussion and concern about the possible hazards of electromagnetic radiation (EMR), including both RF energy and power-frequency (50-60 Hz) electromagnetic fields. FCC regulations set limits on the maximum permissible exposure (MPE) allowed from the operation of radio transmitters.

TOC12 Which of the following steps would help you to comply with RF-radiation exposure guidelines for uncontrolled RF environments?

- A. Reduce transmitting times within a 6-minute period to reduce the station duty cycle
- B. Operate only during periods of high solar absorption
- C. Reduce transmitting times within a 30-minute period to reduce the station duty cycle
- D. Operate only on high duty cycle modes

C   The exposure limits for uncontrolled environments are lower than those for controlled environments, but to compensate for that the standard allows exposure levels in those environments to be averaged over much longer time periods (generally 30 minutes). You have many ways to ensure that no one receives more than the maximum permissible exposure of RF radiation from your station. For example, you can reduce operating power, select an operating mode with a lower duty cycle or even limit your transmit time during any 30-minute averaging period.

TOC13 Which of the following steps would help you to comply with RF-exposure guidelines for controlled RF environments?

- A. Reduce transmitting times within a 30-minute period to reduce the station duty cycle
- B. Operate only during periods of high solar absorption
- C. Reduce transmitting times within a 6-minute period to reduce the station duty cycle
- D. Operate only on high duty cycle modes

C   RF-exposure guidelines for controlled RF environments allow for higher levels of exposure than for uncontrolled areas, but the averaging period is shorter. You can reduce transmitting times within a 6-minute period to reduce the station duty cycle and help you comply with the guidelines for controlled RF environments.

TOC14 To avoid excessively high human exposure to RF fields, how should amateur antennas generally be mounted?

A. With a high current point near ground
B. As far away from accessible areas as possible
C. On a nonmetallic mast
D. With the elements in a horizontal polarization

B   An antenna that is higher and farther away from people reduces the strength of the radiated fields that anyone will be exposed to. If you can raise your antenna higher in the air or move it farther from your neighbor's property line you will reduce exposure. In general, and to reduce human exposure to RF fields, it is better to mount antennas as far away from accessible areas as possible.

TOC15 What action can amateur operators take to prevent exposure to RF radiation in excess of the FCC-specified limits?

A. Alter antenna patterns
B. Relocate antennas
C. Revise station technical parameters, such as frequency, power, or emission type
D. All of these choices are correct

D   If you've read the previous questions in this section, you should realize that all of the choices are correct. Antenna location and pattern are important factors in determining radiated RF fields. So also are transmitter power and emission type. FCC-specified limits are dependent on operating frequency.

TOC16 Which of the following radio frequency emissions will result in the least RF radiation exposure if they all have the same peak envelope power (PEP)?

A. Two-way exchanges of phase-modulated (PM) telephony
B. Two-way exchanges of frequency-modulated (FM) telephony
C. Two-way exchanges of single-sideband (SSB) telephony
D. Two-way exchanges of Morse code (CW) communication

C   Amateur use of SSB results in transmissions with a low duty cycle. When they use FM, PM or RTTY, though, the RF is present continuously at its maximum level during each transmission. CW has a greater duty cycle than SSB but less than FM, PM or RTTY.

TOC17 Why is the concept of "specific absorption rate (SAR)" one factor used to determine safe RF radiation exposure levels?

A. It takes into account the overall efficiency of the final amplifier
B. It takes into account the transmit/receive time ratio during normal amateur communication
C. It takes into account the rate at which the human body absorbs RF energy at a particular frequency
D. It takes into account the antenna feed line loss

C   Specific absorption rate (SAR) is a term that describes the rate at which RF energy is absorbed into the human body. Maximum permissible exposure (MPE) limits are based on whole-body SAR values. This helps explain why these safe exposure limits vary with frequency.

TOC18 Why must the frequency of an RF source be considered when evaluating RF radiation exposure?

A. Lower-frequency RF fields have more energy than higher-frequency fields
B. Lower-frequency RF fields penetrate deeper into the body than higher-frequency fields
C. Higher-frequency RF fields are transient in nature, and do not affect the human body
D. The human body absorbs more RF energy at some frequencies than at others

D   At frequencies near the body's natural resonant frequency, RF energy is absorbed more efficiently, and maximum heating occurs. In adults, this frequency usually is about 35 MHz if the person is grounded, and about 70 MHz if the person's body is insulated from the ground. Also, body parts may be resonant; the adult head, for example is resonant around 400 MHz, while a baby's smaller head resonates near 700 MHz. Body size thus determines the frequency at which most RF energy is absorbed. As the frequency is increased above resonance, less RF heating generally occurs.

TOC19 What is the maximum power density that may be emitted from an amateur station under the FCC RF radiation exposure limits?

    A. The FCC Rules specify a maximum emission of 1.0 milliwatt per square centimeter
    B. The FCC Rules specify a maximum emission of 5.0 milliwatts per square centimeter
    C. The FCC Rules specify exposure limits, not emission limits
    D. The FCC Rules specify maximum emission limits that vary with frequency

C   The FCC Rules specify exposure limits, not emission limits. You can, for example, reduce human exposure to RF fields without reducing emission levels by placing your antenna farther from the places where people live and move.

TOD   **Radio frequency exposure standards; Near/far field, Field strength; Compliance distance; Controlled/ Uncontrolled environment.**

TOD01 What factors must you consider if your repeater station antenna will be located at a site that is occupied by antennas for transmitters in other services?

    A. Your radiated signal must be considered as part of the total RF radiation from the site when determining RF radiation exposure levels
    B. Each individual transmitting station at a multiple transmitter site must meet the RF radiation exposure levels
    C. Each station at a multiple-transmitter site may add no more than 1% of the maximum permissible exposure (MPE) for that site
    D. Amateur stations are categorically excluded from RF radiation exposure evaluation at multiple-transmitter sites

A   If your repeater station antenna will be located at a site that is occupied by antennas for transmitters in other services, your radiated signal must be considered as part of the total RF radiation from the site when determining RF radiation exposure levels. It will help if you remember that FCC is concerned with human exposure to RF. Your repeater cannot be considered alone when there are other transmitters operating from the same location.

T0D02 Why do exposure limits vary with frequency?

    A. Lower-frequency RF fields have more energy than higher-frequency fields

    B. Lower-frequency RF fields penetrate deeper into the body than higher-frequency fields

    C. The body's ability to absorb RF energy varies with frequency

    D. It is impossible to measure specific absorption rates at some frequencies

C   At frequencies near the body's natural resonant frequency, RF energy is absorbed more efficiently, and maximum heating occurs. Body size thus determines the frequency at which most RF energy is absorbed. As the frequency is increased above resonance, less RF heating generally occurs.

T0D03 Why might mobile transceivers produce less RF radiation exposure than hand-held transceivers in mobile operations?

    A. They do not produce less exposure because they usually have higher power levels.

    B. They have a higher duty cycle

    C. When mounted on a metal vehicle roof, mobile antennas are generally well shielded from vehicle occupants

    D. Larger transmitters dissipate heat and energy more readily

C   To reduce RF exposure from a mobile transceiver, you should mount the antenna in the center of the metal roof of your vehicle, if possible. This will use the metal body of the vehicle as an RF shield to protect people inside the car. Glass-mounted antennas can result in higher exposure levels, as can antennas mounted on a trunk lid or front fender. Glass does not form a good RF shield!

TODO4 In the far field, as the distance from the source increases, how does power density vary?

    A. The power density is proportional to the square of the distance

    B. The power density is proportional to the square root of the distance

    C. The power density is proportional to the inverse square of the distance

    D. The power density is proportional to the inverse cube of the distance

    C   Far-field radiation is distinguished by the fact that the power density is proportional to the inverse square of the distance. That means that if you double the distance from the antenna, the power density will be one fourth as strong.

TODO5 In the near field, how does the field strength vary with distance from the source?

    A. It always increases with the cube of the distance

    B. It always decreases with the cube of the distance

    C. It varies as a sine wave with distance

    D. It depends on the type of antenna being used

    D   Nearly any metal object or other conductor that is located within the radiating near field can alter the radiation pattern of the antenna. Even though you may have measured the fields in the general area around your antenna and found that your station meets the MPE limits, there may still be "hot spots" or areas of higher field strengths within that region. In the near field of an antenna, the field strength varies in a way that depends on the type of antenna and other nearby objects as you move farther away from the antenna.

TODO6 Why should you never look into the open end of a microwave feed horn antenna while the transmitter is operating?

A. You may be exposing your eyes to more than the maximum permissible exposure of RF radiation

B. You may be exposing your eyes to more than the maximum permissible exposure level of infrared radiation

C. You may be exposing your eyes to more than the maximum permissible exposure level of ultraviolet radiation

D. All of these choices are correct

A   Microwave emissions are at RF. Infrared and ultraviolet radiation are at higher frequencies. You should never look into the open end of a microwave feed horn antenna while the transmitter is operating. Otherwise, you may be exposing your eyes to more than the maximum permissible exposure of RF radiation.

TODO7 What factors determine the location of the boundary between the near and far fields of an antenna?

A. Wavelength and the physical size of the antenna

B. Antenna height and element length

C. Boom length and element diameter

D. Transmitter power and antenna gain

A   The boundary between the near field and the far field of an antenna is approximately several wavelengths from the antenna. In the near field, ground interactions and other variables produce power densities that cannot be determined by simple arithmetic. In the far field, conditions become easier to predict with simple calculations. It is difficult to accurately evaluate the effects of RF radiation exposure in the near field. The boundary between the near field and the far field depends on the wavelength of the transmitted signal and the physical size and configuration of the antenna.

TODO8 Referring to Figure T0-1, which of the following equations should you use to calculate the maximum permissible exposure (MPE) on the Technician (with code credit) HF bands for a controlled RF radiation exposure environment?

- A. Maximum permissible power density in mW per square cm equals 900 divided by the square of the operating frequency, in MHz
- B. Maximum permissible power density in mW per square cm equals 180 divided by the square of the operating frequency, in MHz
- C. Maximum permissible power density in mW per square cm equals 900 divided by the operating frequency, in MHz
- D. Maximum permissible power density in mW per square cm equals 180 divided by the operating frequency, in MHz

A   As you examine the figure, you'll be looking at the top portion (A). That's the part that covers controlled RF radiation exposure environments. The HF bands are in the frequency range of 3 to 30 MHz. According to the figure, the maximum permissible power density in mw per square cm under these conditions equals 900 divided by the square of the operating frequency, in MHz.

Figure T0- 1

(A) Limits for Occupational/Controlled Exposure				
Frequency Range (MHz)	Electrical Field Strength (V/m)	Magnetic Field Strength (A/m)	Power Density (mW/cm^2)	Averaging Time (minutes)
0.3-3.0	614	1.63	(100)*	6
3.0-30	1842/f	4.89/f	(900/f^2)*	6
30-300	61.4	0.163	1.0	6
300-1500	----	----	f/300	6
1500-100,000	----	----	5	6
(B) Limits for General Population/Uncontrolled Exposure				
Frequency Range (MHz)	Electrical Field Strength (V/m)	Magnetic Field Strength (A/m)	Power Density (mW/cm^2)	Averaging Time (minutes)
0.3-1.34	614	1.63	(100)*	30
1.34-30	824/f	2.19/f	(180/f^2)*	30
30-300	27.5	0.073	0.2	30
300-1500	----	----	f/1500	30
1500-100,000	----	----	1.0	30
f=frequency in MHz *=Plane-wave equivalent power density				

T0D09 Referring to Figure T0-1, what is the formula for calculating the maximum permissible exposure (MPE) limit for uncontrolled environments on the 2-meter (146 MHz) band?

A. There is no formula, MPE is a fixed power density of 1.0 milliwatt per square centimeter averaged over any 6 minutes

B. There is no formula, MPE is a fixed power density of 0.2 milliwatt per square centimeter averaged over any 30 minutes

C. The MPE in milliwatts per square centimeter equals the frequency in megahertz divided by 300 averaged over any 6 minutes

D. The MPE in milliwatts per square centimeter equals the frequency in megahertz divided by 1500 averaged over any 30 minutes

B   Once again refer to the figure. This time look at the bottom (B) part, which is for uncontrolled environments. Find the line labeled 30 – 300 MHz and look under the "Power Density" and "Averaging Time" columns. You find that there is no formula. MPE in this range is a fixed power density of 0.2 milliwatt per square centimeter averaged over any 30 minutes.

T0D10 What is the minimum safe distance for a controlled RF radiation environment from a station using a half-wavelength dipole antenna on 7 MHz at 100 watts PEP, as specified in Figure T0-2?

    A. 1.4 foot
    B. 2 feet
    C. 3.1 feet
    D. 6.5 feet

A   Find the portion of the figure that estimates distances to meet RF power density guidelines with a 7-MHz half-wavelength dipole. On the line for 100 watts PEP transmitter power, find the entry for controlled environments. You'll find the answer is 1.4 feet.

Figure T0-2 (7-MHz Dipole)

Estimated distances to meet RF power density guidelines with a horizontal half-wave dipole antenna (estimated gain, 2 dBi). Calculations include the EPA ground reflection factor of 2.56.

Frequency: 7 MHz
Estimated antenna gain: 2 dBi
Controlled limit: 18.37 mw/cm²
Uncontrolled limit: 3.67 mw/cm²

Transmitter power (watts)	Distance to controlled limit	Distance to uncontrolled limit
100	1.4'	3.1'
500	3.1'	6.9'
1000	4.3'	9.7'
1500	5.3'	11.9'

TOD11 What is the minimum safe distance for an uncontrolled RF radiation environment from a station using a 3-element "triband" Yagi antenna on 28 MHz at 100 watts PEP, as specified in Figure T0-2?

- A. 7 feet
- B. 11 feet
- C. 24.5 feet
- D. 34 feet

C   Find the portion of the figure that estimates distances to meet RF power density guidelines with a 28-MHz "triband" Yagi. On the line for 100 watts PEP transmitter power, check the limit for an uncontrolled RF radiation environment. You'll find the answer is 24.5 feet.

Figure T0-2   (Triband Yagi)

Estimated distances to meet RF power density guidelines in the main beam of a typical 3-element "triband" Yagi for the 14, 21 and 28 MHz amateur radio bands. Calculations include the EPA ground reflection factor of 2.56.

Frequency: 28 MHz
Antenna gain: 8 dBi
Controlled limit: 1.15 mw/cm²
Uncontrolled limit: 0.23 mw/cm²

Transmitter power (watts)	Distance to controlled limit	Distance to uncontrolled limit
100	11'	24.5'
500	24.5'	54.9'
1000	34.7'	77.6'
1500	42.5'	95.1'

T0D12 What is the minimum safe distance for a controlled RF radiation environment from a station using a 146 MHz quarter-wave whip antenna at 10 watts, as specified in Figure T0-2?

A. 1.7 feet
B. 2.5 feet
C. 1.2 feet
D. 2 feet

A   Find the portion of the figure that estimates distances to meet RF power density guidelines with a 146-MHz quarter-wave whip antenna. On the line for 10 watts PEP transmitter power, check the limit for a controlled RF radiation environment. You'll find the answer is 1.7 feet.

Figure T0-2   (146-MHz 1/4-Wavelength Ground Plane)

Estimated distances to meet RF power density guidelines with a VHF quarter-wave ground plane or mobile whip antenna (estimated gain, 1 dBi). Calculations include the EPA ground reflection factor of 2.56.

Frequency: 146 MHz
Estimated antenna gain: 1 dBi
Controlled limit: 1 mw/cm^2
Uncontrolled limit: 0.2 mw/cm^2

Transmitter power (watts)	Distance to controlled limit	Distance to uncontrolled limit
10	1.7'	3.7'
50	3.7'	8.3'
150	6.4'	14.4'

TOD13 What is the minimum safe distance for a controlled RF radiation environment from a station using a 17-element Yagi on a five-wavelength boom on 144 MHz at 100 watts, as specified in Figure T0-2?

- A. 72.4 feet
- B. 78.5 feet
- C. 101 feet
- D. 32.4 feet

D   Find the portion of the figure that estimates distances to meet RF power density guidelines with a 144-MHz 17-element Yagi. On the line for 100 watts PEP transmitter power, check the limit for a controlled RF radiation environment. You'll find the answer is 32.4 feet.

Figure T0-2 (144-MHz Yagi)

Estimated distances to meet RF power density guidelines in the main beam of a 17-element Yagi on a five-wavelength boom designed for weak signal communications on the 144 MHz amateur radio band (estimated gain, 16.8 dBi). Calculations include the EPA ground reflection factor of 2.56.

Frequency: 144 MHz
Estimated antenna gain: 16.8 dBi
Controlled limit: 1 mw/cm^2
Uncontrolled limit: 0.2 mw/cm^2

Transmitter power (watts)	Distance to controlled limit	Distance to uncontrolled limit
10	10.2'	22.9'
100	32.4'	72.4'
500	72.4'	162'
1500	125.5'	280.6'

TOD14 What is the minimum safe distance for an uncontrolled RF radiation environment from a station using a 446 MHz 5/8-wave ground plane vertical antenna at 10 watts, as specified in Figure T0-2?

    A. 1 foot
    B. 4.3 feet
    C. 9.6 feet
    D. 6 feet

B   Find the portion of the figure that estimates distances to meet RF power density guidelines with a 446-MHz 5/8-wave ground plane vertical antenna. On the line for 10 watts PEP transmitter power, check the limit for an uncontrolled RF radiation environment. You'll find the answer is 4.3 feet.

Figure T0-2   (446-MHz 5/8-Wavelength Ground Plane)

Estimated distances to meet RF power density guidelines in the main beam of UHF 5/8 ground plane or mobile whip antenna (estimated gain, 4 dBi). Calculations include the EPA ground reflection factor of 2.56.

Frequency: 446 MHz
Estimated antenna gain: 4 dBi
Controlled limit: 1.49 mw/cm^2
Uncontrolled limit: 0.3 mw/cm^2

Transmitter power (watts)	Distance to controlled limit	Distance to uncontrolled limit
10	1.9'	4.3'
50	4.3'	9.6'
150	7.5'	16.7'

T0E RF Biological effects and potential hazards; Radiation exposure limits; OET Bulletin 65; MPE (Maximum permissible exposure).

T0E01 If you do not have the equipment to measure the RF power densities present at your station, what might you do to ensure compliance with the FCC RF radiation exposure limits?

A. Use one or more of the methods included in the amateur supplement to FCC OET Bulletin 65

B. Call an FCC-Certified Test Technician to perform the measurements for you

C. Reduce power from 200 watts PEP to 100 watts PEP

D. Operate only low-duty-cycle modes such as FM

A Although sophisticated instruments can be used to measure RF power densities accurately, they are costly and require frequent recalibration. Most amateurs don't have access to such equipment, and the inexpensive field-strength meters that we do have are not suitable for measuring RF power density. Fortunately, the FCC has prepared a bulletin, "Amateur Supplement to OET Bulletin 65: Evaluating Compliance With FCC-Specified Guidelines for Human Exposure to Radio Frequency Radiation," that contains charts and tables that amateurs can use to estimate compliance with the rules. A copy of this bulletin is included in the ARRL publication *RF Exposure and You.*

T0E02 Where will you find the applicable FCC RF radiation maximum permissible exposure (MPE) limits defined?

A. FCC Part 97 Amateur Service Rules and Regulations

B. FCC Part 15 Radiation Exposure Rules and Regulations

C. FCC Part 1 and Office of Engineering and Technology (OET) Bulletin 65

D. Environmental Protection Agency Regulation 65

C You'll find the applicable FCC RF radiation maximum permissible exposure (MPE) limits defined in Part 1 of the FCC Rules. The Office of Engineering and Technology (OET) Bulletin 65 also lists the specific MPE limits and provides some information that will help you evaluate your station.

TOE03 To determine compliance with the maximum permitted exposure (MPE) levels, safe exposure levels for RF energy are averaged for an "uncontrolled" RF environment over what time period?

    A. 6 minutes
    B. 10 minutes
    C. 15 minutes
    D. 30 minutes

D   The exposure limits for uncontrolled environments are lower than those for controlled environments, but to compensate for that the standard allows exposure levels in those environments to be averaged over much longer time periods (30 minutes). This longer averaging time means that an intermittently operating RF source (such as an Amateur Radio transmitter) will show a much lower power density than a continuous-duty station for a given power level and antenna configuration.

TOE04 To determine compliance with the maximum permitted exposure (MPE) levels, safe exposure levels for RF energy are averaged for a "controlled" RF environment over what time period?

    A. 6 minutes
    B. 10 minutes
    C. 15 minutes
    D. 30 minutes

A   The exposure limits for controlled environments are higher than those for uncontrolled environments. That is somewhat offset by a shorter averaging period (6 minutes).

TOE05 Why are Amateur Radio operators required to meet the FCC RF radiation exposure limits?

A. The standards are applied equally to all radio services
B. To ensure that RF radiation occurs only in a desired direction
C. Because amateur station operations are more easily adjusted than those of commercial radio services
D. To ensure a safe operating environment for amateurs, their families and neighbors

D   Amateur Radio is basically a safe activity. FCC regulations set limits on the maximum permissible exposure (MPE) allowed from the operation of radio transmitters. These rules have been set in place to ensure a safe operating environment for amateurs, their families and neighbors.

TOE06 At what frequencies do the FCC's RF radiation exposure guidelines incorporate limits for Maximum Permissible Exposure (MPE)?

A. All frequencies below 30 MHz
B. All frequencies between 20,000 Hz and 10 MHz
C. All frequencies between 300 kHz and 100 GHz
D. All frequencies above 300 GHz

C   FCC's RF radiation exposure guidelines cover the radio-frequency spectrum. In this case, that means the frequencies between 300 kHz and 100 GHz.

TOE07 On what value are the maximum permissible exposure (MPE) limits based?

A. The square of the mass of the exposed body
B. The square root of the mass of the exposed body
C. The whole-body specific gravity (WBSG)
D. The whole-body specific absorption rate (SAR)

D   Specific absorption rate (SAR) is a term that describes the rate at which RF energy is absorbed into the human body. Maximum permissible exposure (MPE) limits are based on whole-body SAR values. This helps explain why these safe exposure limits vary with frequency.

TOE08 What is one biological effect to the eye that can result from RF exposure?

A. The strong magnetic fields can cause blurred vision
B. The strong magnetic fields can cause polarization lens
C. It can cause heating, which can result in the formation of cataracts
D. It can cause heating, which can result in astigmatism

C   Anyone who has ever touched an improperly grounded radio chassis and received an RF burn will agree that this type of injury can be quite painful. You do not have to actually touch the chassis to get an RF burn. In extreme cases, RF-induced heating in the eye can result in cataract formation and can even cause blindness.

TOE09 Which of the following effects on the human body are a result of exposure to high levels of RF energy?

A. Very rapid hair growth
B. Very rapid growth of fingernails and toenails
C. Possible heating of body tissue
D. High levels of RF energy have no known effect on the human body

C   At sufficiently high power densities, electromagnetic radiation poses certain health hazards. It has been known since the early days of radio that RF energy can cause injuries by heating body tissue.

TOE10 Why should you not stand within reach of any transmitting antenna when it is being fed with 1500 watts of RF energy?

A. It could result in the loss of the ability to move muscles
B. Your body would reflect the RF energy back to its source
C. It could cause cooling of body tissue
D. You could accidentally touch the antenna and be injured

D   Never stand within reach of a transmitting antenna when it is being fed with high-power RF energy. Anyone who has ever touched an energized antenna and received an RF burn knows that this type of injury can be quite painful.

TOE11 What is one effect of RF non-ionizing radiation on the human body?

A. Cooling of body tissues
B. Heating of body tissues
C. Rapid dehydration
D. Sudden hair loss

B   It has been known since the early days of radio that RF energy can cause injuries by heating body tissue. Serious health problems can result from RF heating. These heat-related health hazards are called thermal effects.

TOF   Routine station evaluation.

TOF01 Is it necessary for you to perform mathematical calculations of the RF radiation exposure if your VHF station delivers more than 50 watts peak envelope power (PEP) to the antenna?

A. Yes, calculations are always required to ensure greatest accuracy
B. Calculations are required if your station is located in a densely populated neighborhood
C. No, calculations may not give accurate results, so measurements are always required
D. No, there are alternate means to determine if your station meets the RF radiation exposure limits

D   If you run more than 50 watts peak envelope power (PEP) to the antenna at VHF, you are required to perform a routine RF radiation evaluation. The FCC maximum permissible exposure limits are given in terms of electric and magnetic field strengths. So how do you determine if the transmitted signal from your station is within these RF exposure limits? You must analyze, measure or otherwise determine your transmitted field strengths and power density. This means there are a number of ways you can perform the required routine RF radiation evaluation. Depending on the method you choose, you may not have to perform any mathematical calculations to do an evaluation.

TOF02 What is one method that amateur radio licensees may use to conduct a routine station evaluation to determine whether the station is within the Maximum Permissible Exposure guidelines?

    A. Direct measurement of the RF fields

    B. Indirect measurement of the energy density at the limit of the controlled area

    C. Estimation of field strength by S-meter readings in the controlled area

    D. Estimation of field strength by taking measurements using a directional coupler in the transmission line

A   One way to conduct a routine station evaluation is by making direct measurements of the electric and magnetic field strengths around your antenna while transmitting a signal. If you happen to have a calibrated field-strength meter with a calibrated field-strength sensor, you can make accurate measurements. Unfortunately, such calibrated meters are expensive and not normally found in an amateur's toolbox. The relative field strength meters many amateurs use are not accurate enough to make this type of measurement.

TOF03 What document establishes mandatory procedures for evaluating compliance with RF exposure limits?

    A. There are no mandatory procedures

    B. OST/OET Bulletin 65

    C. Part 97 of the FCC rules

    D. ANSI/IEEE C95.1—1992

A   This may surprise you, but there are no mandatory procedures for performing a compliance evaluation. You are required to evaluate your station's compliance with RF exposure limits. There are different methods that you can use, and the choice of them is up to you.

TOF04 Which category of transceiver is NOT excluded from the requirement to perform a routine station evaluation?

A. Hand-held transceivers
B. VHF base station transmitters that deliver more than 50 watts peak envelope power (PEP) to an antenna
C. Vehicle-mounted push-to-talk mobile radios
D. Portable transceivers with high duty cycles

B   Amateur mobile and portable hand-held stations using push-to-talk or equivalent operation are not required to perform routine station evaluations. (These stations are *not* exempt from the rules, but are presumed to be in compliance without the need for an evaluation.) VHF base station transmitters that deliver more than 50 watts peak envelope power (PEP) to an antenna are required to perform a routine station evaluation.

TOF05 Which of the following antennas would (generally) create a stronger RF field on the ground beneath the antenna?

A. A horizontal loop at 30 meters above ground
B. A 3-element Yagi at 30 meters above ground
C. A 1/2 wave dipole antenna 5 meters above ground
D. A 3-element Quad at 30 meters above ground

C   A half-wavelength dipole antenna that is only 5 meters above the ground would generally create a stronger RF field on the ground beneath the antenna than many other antennas. For example, a horizontal loop, a 3-element Yagi antenna or a 3-element Quad antenna all have significantly more gain than a dipole. Yet at a height of 30 meters each of these antennas would produce a smaller RF field strength on the ground beneath the antenna than would the low dipole. As a general rule, you should place your antenna at least as high as necessary to ensure that you meet the FCC radiation exposure guidelines.

TOF06 How may an amateur determine that his or her station complies with FCC RF-exposure regulations?

A. By calculation, based on FCC OET Bulletin No. 65
B. By calculation, based on computer modeling
C. By measurement, measuring the field strength using calibrated equipment
D. Any of these choices

D   You may use a variety of methods to determine that your station complies with FCC RF-exposure regulations. All of the choices given above are correct and valid methods of making that determination.

TOF07 Below what power level at the input to the antenna are amateur radio operators categorically excluded from routine evaluation to predict if the RF exposure from their VHF station could be excessive?

A. 25 watts peak envelope power (PEP)
B. 50 watts peak envelope power (PEP)
C. 100 watts peak envelope power (PEP)
D. 500 watts peak envelope power (PEP)

B   If you are running less than 50 watts peak envelope power (PEP) at VHF to the input of your antenna, you are categorically excluded from routine RF exposure evaluation.

TOF08 Above what power level is a routine RF radiation evaluation required for a VHF station?

A. 25 watts peak envelope power (PEP) measured at the antenna input
B. 50 watts peak envelope power (PEP) measured at the antenna input
C. 100 watts input power to the final amplifier stage
D. 250 watts output power from the final amplifier stage

B   If you are running 50 watts peak envelope power (PEP) or more at VHF to the input of your antenna, you are required to perform a routine RF exposure evaluation of your station.

TOF09 What must you do with the records of a routine RF radiation exposure evaluation?

    A. They must be sent to the nearest FCC field office

    B. They must be sent to the Environmental Protection Agency

    C. They must be attached to each Form 605 when it is sent to the FCC for processing

    D. Though not required, records may prove useful if the FCC asks for documentation to substantiate that an evaluation has been performed

D   Your are not *required* to keep a record of your routine RF radiation exposure evaluation(s). Nevertheless, it is a *good idea* to keep those records in a handy location. They could prove useful if the FCC asks for documentation to substantiate that an evaluation has been performed.

TOF10 Which of the following instruments might you use to measure the RF radiation exposure levels in the vicinity of your station?

    A. A calibrated field strength meter with a calibrated field strength sensor

    B. A calibrated in-line wattmeter with a calibrated length of feed line

    C. A calibrated RF impedance bridge

    D. An amateur receiver with an S meter calibrated to National Bureau of Standards and Technology station WWV

A   You could use a calibrated field strength meter with a calibrated field strength sensor to measure the RF radiation exposure levels in the vicinity of your station. The items listed in the other choices are not capable of making the necessary accurate field strength measurements.

TOF11 What effect does the antenna gain have on a routine RF exposure evaluation?

    A. Antenna gain is part of the formulas used to perform calculations

    B. The maximum permissible exposure (MPE) limits are directly proportional to antenna gain

    C. The maximum permissible exposure (MPE) limits are the same in all locations surrounding an antenna.

    D. All of these choices are correct

A   You can perform some calculations to determine the electric and magnetic field strengths from your station. These calculations take into account the gain and directivity of an antenna. In other words, antenna gain is part of the formulas used to perform calculations. That's true whether you're doing manual calculations or using a computer program to assist you.

TOF12 As a general rule, what effect does antenna height above ground have on the RF exposure environment?

    A. Power density is not related to antenna height or distance from the RF exposure environment

    B. Antennas that are farther above ground produce higher maximum permissible exposures (MPE)

    C. The higher the antenna the less the RF radiation exposure at ground level

    D. RF radiation exposure is increased when the antenna is higher above ground

C   As a general rule, the farther you are from an antenna, the lower your RF radiation exposure. In the far field, power density is proportional to the inverse square of the distance. That means if you double the distance from the antenna, the power density will be one fourth as strong. All this also means that the higher the antenna the less the RF radiation exposure at ground level.

TOF13 Why does the FCC consider a hand-held transceiver to be a portable device when evaluating for RF radiation exposure?

A. Because it is generally a low-power device
B. Because it is designed to be carried close to your body
C. Because it's transmitting antenna is generally within 20 centimeters of the human body
D. All of these choices are correct

C   Hand-held radios are very popular for VHF and UHF operation, especially with FM repeaters. They transmit with less than 7-watts of power, which is generally considered to be safe. Because the radios are designed to be operated with an antenna that is within 20 centimeters of your body, they are classified as portable devices by the FCC.

TOF14 Which of the following factors must be taken into account when using a computer program to model RF fields at your station?

A. Height above sea level at your station
B. Ionization level in the F2 region of the ionosphere
C. Ground interactions
D. The latitude and longitude of your station location

C   Computer antenna-modeling programs such as *MININEC* or other codes derived from *NEC* (Numerical Electromagnetic Code) are suitable for estimating RF magnetic and electric fields around amateur antenna systems. These programs, however, have their limitations. Ground interactions must be considered in estimating near-field power densities. Also, computer modeling is not sophisticated enough to predict "hot spots" in the near field—places where the field intensity may be far higher than would be expected.

TOF15 In which of the following areas is it most difficult to accurately evaluate the effects of RF radiation exposure?

A. In the far field
B. In the cybersphere
C. In the near field
D. In the low-power field

C   It is possible to calculate the probable power density near an antenna using relatively simple equations. However, such calculations have many pitfalls. For one, most of the situations in which the power density would be high enough to be of concern are in the near field. In the near field, ground interactions and other variables produce power densities that cannot be determined by simple arithmetic. In the far field, conditions become easier to predict with simple calculations.

About the ARRL _____

The seed for Amateur Radio was planted in the 1890s, when Guglielmo Marconi began his experiments in wireless telegraphy. Soon he was joined by dozens, then hundreds, of others who were enthusiastic about sending and receiving messages through the air—some with a commercial interest, but others solely out of a love for this new communications medium. The United States government began licensing Amateur Radio operators in 1912.

By 1914, there were thousands of Amateur Radio operators—hams—in the United States. Hiram Percy Maxim, a leading Hartford, Connecticut inventor and industrialist, saw the need for an organization to band together this fledgling group of radio experimenters. In May 1914 he founded the American Radio Relay League (ARRL) to meet that need.

Today ARRL, with approximately 170,000 members, is the largest organization of radio amateurs in the United States. The ARRL is a not-for-profit organization that:

- promotes interest in Amateur Radio communications and experimentation
- represents US radio amateurs in legislative matters, and
- maintains fraternalism and a high standard of conduct among Amateur Radio operators.

At ARRL headquarters in the Hartford suburb of Newington, the staff helps serve the needs of members. ARRL is also International Secretariat for the International Amateur Radio Union, which is made up of similar societies in 150 countries around the world.

ARRL publishes the monthly journal *QST*, as well as newsletters and many publications covering all aspects of Amateur Radio. Its headquarters station, W1AW, transmits bulletins of interest to radio amateurs and Morse code practice sessions. The ARRL also coordinates an extensive field organization, which includes volunteers who provide technical information and other support services for radio amateurs as well as communications for public-service activities. In addition, ARRL represents US amateurs with the Federal Communications Commission and other government agencies in the US and abroad.

Membership in ARRL means much more than receiving *QST* each month. In addition to the services already described, ARRL offers membership services on a personal level, such as the ARRL Volunteer Examiner Coordinator Program and a QSL bureau.

Full ARRL membership (available only to licensed radio amateurs) gives you a voice in how the affairs of the organization are governed.

ARRL policy is set by a Board of Directors (one from each of 15 Divisions). Each year, one-third of the ARRL Board of Directors stands for election by the full members they represent. The day-to-day operation of ARRL HQ is managed by an Executive Vice President and his staff.

No matter what aspect of Amateur Radio attracts you, ARRL membership is relevant and important. There would be no Amateur Radio as we know it today were it not for the ARRL. We would be happy to welcome you as a member! (An Amateur Radio license is not required for Associate Membership.) For more information about ARRL and answers to any questions you may have about Amateur Radio, write or call:

ARRL—The national association for Amateur Radio
225 Main Street
Newington CT 06111-1494

Voice: 860-594-0200
Fax: 860-594-0259

E-mail: **hq@arrl.org**
Internet: **www.arrl.org/**

Prospective new amateurs call (toll-free):
800-32-NEW HAM (800-326-3942)
You can also contact us via e-mail at **newham@arrl.org**
or check out **ARRLWeb** at **http://www.arrl.org/**

JOIN ARRL TODAY AND RECEIVE A *FREE* BOOK!

I want to join ARRL. Send me the *FREE* book I have selected (choose one):

☐ *Repeater Directory*—Gives you listings of more than 20,000 FM voice repeaters throughout the US.

☐ *The Best of the New Ham Companion*—Selected articles on all aspects of ham radio—from the beginner's perspective.

☐ New member   ☐ Previous member   ☐ Renewal

_____
Call Sign (if any)   Class of License   Date of Birth

_____
Name

_____
Address

_____
City, State, ZIP

Dues are $39* in the US. You do not need an Amateur Radio license to join. Individuals age 65 or over, residing in the US, upon submitting one-time date of birth, may request the dues rate of $34*. Immediate relatives of a member who receives *QST*, and reside at the same address may request family membership at $8 per year. Blind individuals may join without *QST* for $8 per year. If you are 21 or younger and a licensed amateur, a special rate may apply. Write or call ARRL for details.

Sorry! Free book offer does not apply to individuals joining as family or blind members or submitting their application via clubs.

*One-year membership includes $15 for a one-year subscription to *QST*. Memberships and *QST* cannot be separated.

DUES ARE SUBJECT TO CHANGE WITHOUT NOTICE.

Amount Enclosed-Payable to ARRL $_____

Charge to MC, VISA, AMEX, Discover No. _____

Expiration Date_____

Cardholder Name_____

Cardholder Signature_____

If you do not wish your name and address made available for non-ARRL related mailings, please check this box ☐

ARRL
225 MAIN STREET NEWINGTON, CONNECTICUT 06111 USA
New Hams call (800) 326-3942
Call toll free to join: (888) 277-5289
Join on the Web: http://www.arrl.org/join.html

TQA 03

245

FEEDBACK

Please use this form to give us your comments on this book and what you'd like to see in future editions, or e-mail us at **pubsfdbk@arrl.org** (publications feedback). If you use e-mail, please include your name, call, e-mail address and the book title, edition and printing in the body of your message. Also indicate whether or not you are an ARRL member.

Please check the box that best answers these questions:
How well did this book prepare you for your exam?
    ☐ Very Well   ☐ Fairly Well   ☐ Not Very Well
Which exam did you take (or will you be taking)?
    ☐ Technician   ☐ Technician with code   ☐ General
Did you pass?   ☐ Yes   ☐ No
Do you expect to learn Morse code some time? ☐ Yes ☐ No ☐ Already know code
Where did you purchase this book?
    ☐ From ARRL directly   ☐ From an ARRL dealer

Is there a dealer who carries ARRL publications within:
☐ 5 miles   ☐ 15 miles   ☐ 30 miles of your location? ☐ Not sure.

If licensed, what is your license class? _____

Name _____ ARRL member? ☐ Yes ☐ No
_____ Call Sign_____
Address _____
City, State/Province, ZIP/Postal Code _____
Daytime Phone ( ) _____ Age _____ E-mail _____
If licensed, how long?_____
Other hobbies_____

Occupation _____

For ARRL use only	T Q&A
Edition	3 4 5 6 7 8 9 10 11
Printing	2 3 4 5 6 7 8 9 10 11

From _____

_____

_____

EDITOR, THE ARRL'S TECH Q&A
ARRL—THE NATIONAL ASSOCIATION FOR AMATEUR RADIO
225 MAIN STREET
NEWINGTON CT 06111-1494

....:please fold and tape...